Continuum Mechanics - Progress in Fundamentals and Engineering Applications

Continuum Mechanics - Progress in Fundamentals and Engineering Applications

Editor

Edvin Falk

scitus
academics

Continuum Mechanics - Progress in Fundamentals and Engineering Applications
Edited by **Edvin Falk**

ISBN: 978-1-68117-286-6
Library of Congress Control Number: 2016937462

SCITUS Academics LLC,
www.scitusacademics.com
Box No. 4766, 616 Corporate Way,
Suite 2, Valley Cottage,
NY 10989

Notice

Reasonable efforts have been made to publish reliable data and views articulated in the chapters are those of the individual contributors, and not necessarily those of the editors or publishers. Editors or publishers are not responsible for the accuracy of the information in the published chapters or consequences of their use. The publisher believes no responsibility for any damage or grievance to the persons or property arising out of the use of any materials, instructions, methods or thoughts in the book. The editors and the publisher have attempted to trace the copyright holders of all material reproduced in this publication and apologize to copyright holders if permission has not been obtained. If any copyright holder has not been acknowledged, please write to us so we may rectify.

Printed in United States of America on Acid Free Paper

Preface

Materials such as solids, liquids and gasses are composed of molecules separated by 'empty' space. Microscopic materials have cracks and discontinuities. However, certain physical phenomena can be modeled on the assumption that the materials exist as a continuum, which means that the matter is continually distributed throughout the body and fills the entire space region. A continuum is a body which can be continually divided into infinitesimal elements with the properties of the material in bulk. Continuum mechanics, a mechanical branch, deals with the analysis of cinematics and the mechanical behavior of materials modeled instead of discrete particles as a continuous mass. Continuum mechanics compacts with the physical properties of solids and fluids in which they are observed independently of any specific coordinate system. These physical properties are represented by tensors, which are mathematical objects with the property to be independent of the system of coordinates. These tensors can be expressed in co-ordinated computer systems.

This book examines the progress of continuum mechanics in a number of defined areas. The book emphasizes the application aspects of the subject. It clearly demonstrates general principles common to solid and fluid mechanics. This subject also talks about the need for solid and fluid assumptions—in the form of component equations. The mechanical continuum frame work is also useful to understand elasticity, plasticity, viscoelasticity, and viscoplasticity.

Table of Contents

CHAPTER 1

Efficient Simulation of Nonlinear Heat Transfer during Thermal Spraying of Complex Workpieces

Rolf Berthelsen[1*], Thomas Wiederkehr[2], Ralf Denzer[1], Andreas Menzel[1,3], Heinrich Müller[2]

[1]Institute of Mechanics, TU Dortmund, Leonhard-Euler-Straße 5, Dortmund, Germany

[2]Lehrstuhl Informatik VII, TU Dortmund, Otto-Hahn-Straße 16, Dortmund, Germany

[3]Division of Solid Mechanics, Lund University, Ole Römers väg, Lund, Sweden

ABSTRACT

The quality of coatings, produced by thermal spraying processes, considerably decreases with the occurrence of higher residual stresses, which are especially pronounced for complex workpiece geometries. To understand the occurring effects and to aid in the planning of coating processes, simulations of the highly transient energy flux of the HVOF spray gun into the substrate are of great value. In this article, a software framework for the simulation of nonlinear heat transfer during (HVOF) thermal spraying is presented. One part of this framework employs an efficient GPU-based simulation algorithm to compute the time-dependent input boundary conditions for a spray gun that moves along a complex workpiece of arbitrary shape. The other part employs a finite-element model for a rigid heat conductor adhering to the computed boundary conditions. The model is derived from the fundamental equations of continuum thermodynamics where nonlinear temperature-depending heat conduction is assumed.

KEYWORDS

Nonlinear Heat Conduction, Continuum Thermodynamics, GPU Computing

1. INTRODUCTION

Thermal spraying is a cost-efficient coating technique for the production of wear-resistant surfaces consisting of various materials tailored to particular applications. For the coating of forming tools, e.g., hard material coatings of tungsten carbide (WC) and cobalt (Co) are used, [1] , because of their superior wear-resistance compared to chrome (Cr) and nickel (Ni) coatings, [2] . As a disadvantage, the High Velocity Oxygen Fuel (HVOF) thermal spraying process induces a large amount of energy into the heterogeneous coating and the substrate which leads to a complex transient thermo-mechanical problem, as illustrated in Figure 1. For an overview about HVOF thermal spraying, the reader is referred to, e.g., [3] [4] . While thermal-spraying of planar work pieces delivers coatings of quite satisfying quality, the quality considerably decreases with the rising degree of complexity of the work pieces' geometry, for instance due to radii or curvatures. Experimental analyses suggest significant temperature exaltations in the substrate in dependence of the component geometry, whereby the coating quality decreases due to residual stresses during the cooling process. Thus, for the computer aided planning of thermal spray processes, the ability to predict the temperature development for a given workpiece is of high value.

This article presents a simulation framework for the computation of the temperature development for a given workpiece during the thermal spraying process. Therein the "outer" part of the framework handles the time stepping process, the spray gun movement and computes the time-dependent input boundary conditions for the "inner" simulation module, which evaluates a thermodynamically consistent transient and nonlinear heat conduction formulation for a rigid heat conductor based on the governing equations of continuum thermodynamics, as introduced by Coleman and Noll [5] .

For complex workpieces the setup of the input boundary conditions is a computationally expensive procedure by itself, since the (non-ambient) load distribution on the workpiece surface due to thermal radiation is not locally restricted and may affect a large fraction of the workpiece. The computation of the affected parts constitutes a computationally expensive visibility problem. For this, a novel approach is presented, which makes use of an elegant GPU-acceleration technique and is in fact an adaption of a method developed earlier in the context of mass deposition simulation [6] . This two-level simulation approach enables the highly detailed thermodynamic model, which is specifically adjusted to the HVOF spraying process, to be efficiently applied even to larger workpieces such as forming tools in the automotive industry or complex geometries like turbine blades. Another contribution of this work is the "inner" finite element simulation module including a multi-threaded C++ based implementation which uses the solver library Eigen.

The following two sections describe the fundamental equations of continuum thermodynamics for a rigid heat conductor and the resulting finite element discretisation. Section 4 outlines the functionality of the robot guided spray simulation, [6] , and provides important details regarding the GPU-accelerated computation of the input boundary conditions. Section 5 presents a demonstration of the software tool for the simulation of a real work piece. The material parameters of the underlying constitutive relations—the heat capacity c as well as the heat conduction coefficient λ —are represented by suitable functions which are fitted to experimental data.

Figure 1. Photograph of the HVOF spraying process, kindly provided by LWT, TU Dortmund.

2. CONTINUUM THERMODYNAMICS

The governing well-established equations of continuum thermodynamics for a rigid heat conductor are illustrated in the following section. The fundamental balance equation for a rigid heat conductor is the balance of energy, here in local form,

$$\rho\dot{e}+\nabla_X \cdot \boldsymbol{q} = q, \tag{1}$$

wherein $\dot{e}$ denotes the rate of the internal energy, $\boldsymbol{q}$ denotes the heat flux vector and q denotes the external heat supply, while ρ represents the material density, [5] . The balance of energy equation represents the first law of thermodynamics. By assuming the entropy flux to be proportional to the heat flux, as proposed by Coleman and Noll [5] , the general entropy inequality results in the so called Clausius-Duhem inequality

$$\rho\dot{\eta}+\nabla_X \cdot \frac{\boldsymbol{q}}{\theta} - \frac{q}{\theta} \geq 0. \tag{2}$$

In the above inequality, $\dot{\eta}$ is the rate of the entropy and θ is the absolute temperature. The Clausius- Duhem inequality represents the second law of thermodynamics. A Legendre transform of the internal energy $e(\eta)$ results in the free energy density $\psi(\theta) = e - \theta\eta$ so that $\eta = -\partial\psi/\partial\theta$ and subsequently $e = \psi - \theta\partial\psi/\partial\theta$. Following Liu [7] and references cited therein, the only remaining part of inequality

(2) is the thermal dissi- pation $\mathcal{D}_{\text{therm}} = -\boldsymbol{q}\cdot\nabla_X\theta/\theta \geq 0$, also referred to as Fourier's inequality which is fulfilled if the heat flux points from a hot spot to a cold spot. To maintain generality, the specific heat is introduced temperature-dependent, i.e. $c(\theta) = \partial e/\partial\theta = -\theta\left(\partial^2\psi\right)/\left(\partial\theta^2\right)$. The energy balance (1) is now restated in the strong form of the temperature field equation valid for a general rigid heat conductor,

$$\rho c(\theta)\dot{\theta} + \nabla_X \cdot \boldsymbol{q} = q. \tag{3}$$

This equation is to be solved within a general thermodynamical initial boundary value problem. For this pur- pose, the boundary $\partial\mathcal{B}$ of the body $\mathcal{B}$ is decomposed into three disjoint parts, i.e., $\partial\mathcal{B}^{\mathrm{D}} \cup \partial\mathcal{B}^{\mathrm{N}} \cup \partial\mathcal{B}^{\mathrm{R}} = \partial\mathcal{B}$ with $\partial\mathcal{B}^{\mathrm{D}} \cap \partial\mathcal{B}^{\mathrm{N}} = \varnothing$, $\partial\mathcal{B}^{\mathrm{D}} \cap \partial\mathcal{B}^{\mathrm{R}} = \varnothing$ and $\partial\mathcal{B}^{\mathrm{N}} \cap \partial\mathcal{B}^{\mathrm{R}} = \varnothing$. On $\partial\mathcal{B}^{\mathrm{D}}$ Dirichlet boundary conditions are prescribed for the temperature θ, whereas Neumann and Robin boundary conditions are prescribed for the heat flux $\boldsymbol{q}$ on $\partial\mathcal{B}^{\mathrm{N}}$ respectively $\partial\mathcal{B}^{\mathrm{R}}$:

$$\theta = \bar{\theta} \text{ on } \partial\mathcal{B}^{\mathrm{D}}, \quad -\boldsymbol{q}\cdot\boldsymbol{n} = \bar{q}^{\mathrm{N}} \text{ on } \partial\mathcal{B}^{\mathrm{N}}, \quad -\boldsymbol{q}\cdot\boldsymbol{n} = \bar{q}^{\mathrm{R}}(\theta) \text{ on } \partial\mathcal{B}^{\mathrm{R}}. \tag{4}$$

Fourier's law of heat conduction is applied throughout this work. For a review on non-Fourier heat conduction the reader is referred to other recent works, e.g., the works of Atefi and Talaee [8] [9] .

3. FINITE ELEMENT DISCRETISATION

The different representations of the energy balance equation derived above are given in strong form. To calculate a solution for the desired field of the temperature by means of the finite element method in the context of inhomogeneous initial boundary value problems, the temperature-based balance of energy has to be reformulated in weak form. Therefore, we transfer Equation (3) to a residual form,

$$z(\omega;\theta) = \omega\left[\rho c\dot{\theta} + \nabla_X \cdot \boldsymbol{q} - q\right] = 0, \tag{5}$$

where ω is a scalar-valued test function, see [10] for detailed background information. This relation also holds under integration over the volume of a body $\mathcal{B}$,

$$Z(\omega;\theta) = \int_B \omega\left[\rho c\dot{\theta} + \nabla_X \cdot \boldsymbol{q} - q\right] \mathrm{d}V = 0. \tag{6}$$

The divergence of the heat flux in Equation (6) can be reformulated by the application of Gauß's theorem and integration by parts,

$$\int_B \omega\rho c\dot{\theta}\mathrm{d}V - \int_B \omega q\mathrm{d}V - \int_B \nabla_X \omega \cdot \boldsymbol{q}\mathrm{d}V + \int_{\partial B} \omega\boldsymbol{q} \cdot \boldsymbol{n}\mathrm{d}A = 0. \tag{7}$$

Here, $\boldsymbol{n}$ denotes the outward unit vector on the boundary of the body $\mathcal{B}$. By interpreting the test function ω as the virtual temperature $\delta\theta$, which is 0 at Dirichlet boundaries $\partial\mathcal{B}^{\mathrm{D}}$, Equation (7) represents the virtual temperature problem and can be written in terms of dynamic, volume, internal and surface terms

$$g(\delta\theta;\theta) = w_{\mathrm{dyn}} - w_{\mathrm{vol}} - w_{\mathrm{int}} + w_{\mathrm{sur}} = 0 \quad \forall \delta\theta, \tag{8}$$

which, in turn, are defined by

$$w_{\mathrm{dyn}} = \int_{\mathcal{B}} \delta\theta \rho c \dot{\theta} \mathrm{d}V, \quad w_{\mathrm{vol}} = \int_{\mathcal{B}} \delta\theta q \mathrm{d}V, \quad w_{\mathrm{int}} = \int_{\mathcal{B}} \nabla_X \delta\theta \cdot \boldsymbol{q} \mathrm{d}V, \quad w_{\mathrm{sur}} = \int_{\mathcal{B}} \delta\theta \boldsymbol{q} \cdot \boldsymbol{n} \mathrm{d}A. \tag{9}$$

Equation (8) is the weak form of the initial boundary value problem of a rigid heat conductor which now has to be discretised in time and space, following the procedure outlined, e.g., in [11] or Kuhl et al. [12] . For time discretisation, a differential-quotient-based Backward Euler integration scheme is applied [13] ,

$$\frac{\partial[\bullet]}{\partial t} \approx \frac{[\bullet]_{n+1} - [\bullet]_n}{\Delta t}, \tag{10}$$

where $[\bullet]$ symbolises an arbitrary quantity of interest and Δt denotes the time increment $t_{n+1} - t_n$. Subscript $n+1$ denotes a quantity at the actual time step t_{n+1}, whereas subscript n denotes a quantity at the previous time step t_n. Application of Equation (10) to the temperature θ results in

$$\dot{\theta} = \frac{\theta_{n+1} - \theta_n}{\Delta t}. \tag{11}$$

In addition to the Backward Euler method applied in this work, recent works which address thermal problems use the Crank-Nicolson method [14] or energy-momentum consistent schemes in the context of thermo- elastodynamics [15] . At time t_{n+1}, Equation (11) leads to the relation

$$g(\delta\theta; \theta_{n+1}) = w_{\text{dyn}} - w_{\text{vol}} - w_{\text{int}} + w_{\text{sur}} = 0 \quad \forall \delta\theta, \tag{12}$$

for the unknown temperature θ_{n+1}. In view of the discretisation in space, it has to be taken into account that Robin boundary conditions require additional effort, since they represent one type of temperature dependent loads. For a review on the implementation and algorithmic treatment of deformation dependent loads, the reader is referred to the textbook by Bonet and Wood [16] and references cited therein. For that purpose, on the one hand the body

$\mathcal{B}$ is approximated by a finite number of $n_{el}^{\mathcal{B}}$ volume elements $\mathcal{B}_{\mathcal{B}}^{e}$ and, on the other hand, the Robin boundary $\partial\mathcal{B}^{\mathrm{R}}$ of the body $\mathcal{B}$ is approximated by $n_{el}^{\partial\mathcal{B}}$ surface elements $\mathcal{B}_{\partial\mathcal{B}}^{e}$, i.e.

$$\mathcal{B} \approx \mathcal{B}^{h} = \bigcup_{e=1}^{n_{el}^{\mathcal{B}}} \mathcal{B}_{\mathcal{B}}^{e}, \quad \partial\mathcal{B}^{\mathrm{R}} \approx \partial\mathcal{B}^{\mathrm{R}h} = \bigcup_{e=1}^{n_{el}^{\partial\mathcal{B}}} \mathcal{B}_{\partial\mathcal{B}}^{e}. \tag{13}$$

Following the spirit of the isoparametric concept, the geometry of the body, in terms of position vectors $\boldsymbol{X}$, as well as the temperature of the body θ and the virtual temperature $\delta\theta$ are interpolated element-wise by shape functions $N_{X,\mathcal{B}}^{i}$ respectively $N_{X,\partial\mathcal{B}}^{i}$ and discrete node point positions $\boldsymbol{X}_i$ of $i=1,\cdots,n_{en}^{\mathcal{B}}$ respectively $i=1,\cdots,n_{en}^{\partial\mathcal{B}}$ element nodes, i.e.

$$\begin{aligned}
\boldsymbol{X}_{\mathcal{B}} &\approx \boldsymbol{X}_{\mathcal{B}}^{h} = \bigcup_{i=1}^{n_{en}^{\mathcal{B}}} N_{X,\mathcal{B}}^{i} \boldsymbol{X}_i, & \boldsymbol{X}_{\partial\mathcal{B}} &\approx \boldsymbol{X}_{\partial\mathcal{B}}^{h} = \bigcup_{i=1}^{n_{en}^{\partial\mathcal{B}}} N_{X,\partial\mathcal{B}}^{i} \boldsymbol{X}_i, \\
\theta_{\mathcal{B}} &\approx \theta_{\mathcal{B}}^{h} = \bigcup_{i=1}^{n_{en}^{\mathcal{B}}} N_{X,\mathcal{B}}^{i} \theta_i, & \theta_{\partial\mathcal{B}} &\approx \theta_{\partial\mathcal{B}}^{h} = \bigcup_{i=1}^{n_{en}^{\partial\mathcal{B}}} N_{X,\partial\mathcal{B}}^{i} \theta_i, \\
\delta\theta_{\mathcal{B}} &\approx \delta\theta_{\mathcal{B}}^{h} = \bigcup_{i=1}^{n_{en}^{\mathcal{B}}} N_{X,\mathcal{B}}^{i} \delta\theta_i, & \delta\theta_{\partial\mathcal{B}} &\approx \delta\theta_{\partial\mathcal{B}}^{h} = \bigcup_{i=1}^{n_{en}^{\partial\mathcal{B}}} N_{X,\partial\mathcal{B}}^{i} \delta\theta_i.
\end{aligned} \tag{14}$$

Hence, the gradients $\nabla_X \boldsymbol{X}$, $\nabla_X \theta$ and $\nabla_X \delta\theta$ are approximated as

$$
\begin{aligned}
&\nabla_X \boldsymbol{X}_B \approx \nabla_X \boldsymbol{X}_B^h = \bigcup_{i=1}^{n_{en}^B} \boldsymbol{X}_i \otimes \nabla_X N_{X,B}^i, && \nabla_X \boldsymbol{X}_{\partial B} \approx \nabla_X \boldsymbol{X}_{\partial B}^h = \bigcup_{i=1}^{n_{en}^{\partial B}} \boldsymbol{X}_i \otimes \nabla_X N_{X,\partial B}^i, \\
&\nabla_X \theta_B \approx \nabla_X \theta_B^h = \bigcup_{i=1}^{n_{en}^B} \theta_i \nabla_X N_{X,B}^i, && \nabla_X \theta_{\partial B} \approx \nabla_X \theta_{\partial B}^h = \bigcup_{i=1}^{n_{en}^{\partial B}} \theta_i \nabla_X N_{X,\partial B}^i, \\
&\nabla_X \delta\theta_B \approx \nabla_X \delta\theta_B^h = \bigcup_{i=1}^{n_{en}^B} \delta\theta_i \nabla_X N_{X,B}^i, && \nabla_X \delta\theta_{\partial B} \approx \nabla_X \delta\theta_{\partial B}^h = \bigcup_{i=1}^{n_{en}^{\partial B}} \delta\theta_i \nabla_X N_{X,\partial B}^i.
\end{aligned} \tag{15}
$$

The insertion of the approximations given by Equations (14) and (15) into Equation (12) leads to the fully discretised balance of energy in terms of the unknown temperature θ_{n+1}

$$
r_I^h\left(\theta_{n+1}^h\right) = f_{\mathrm{dyn}I} - f_{\mathrm{vol}I} - f_{\mathrm{int}\,I} + f_{\mathrm{sur}I} = 0, \tag{16}
$$

where the discrete inertia, volume, internal and surface fluxes are expressed by

$$
\begin{aligned}
f_{\mathrm{dyn}I} &= \mathop{\mathbf{A}}_{e=1}^{n_{el}^B} \int_{B_B^e} N_{X,B}^i c(\theta) \frac{\theta_{n+1} - \theta_n}{\Delta t} \mathrm{d}V, \\
f_{\mathrm{vol}I} &= \mathop{\mathbf{A}}_{e=1}^{n_{el}^B} \int_{B_B^e} N_{X,B}^i q \mathrm{d}V, \\
f_{\mathrm{int}\,I} &= \mathop{\mathbf{A}}_{e=1}^{n_{el}^B} \int_{B_B^e} \nabla_X N_{X,B}^i \cdot \boldsymbol{q} \mathrm{d}V, \\
f_{\mathrm{sur}I} &= \mathop{\mathbf{A}}_{e=1}^{n_{el}^B} \int_{\partial B_B^e} -N_{X,B}^i \bar{q}^{\mathrm{N}} \mathrm{d}A + \mathop{\mathbf{A}}_{e=1}^{n_{el}^B} \int_{B_{\partial B}^e} -N_{X,B}^i \bar{q}^{\mathrm{R}}(\theta) \mathrm{d}A.
\end{aligned} \tag{17}
$$

In the above equations, **A** represents the assembly operator over all $e=1,\cdots,n_{el}^{\mathcal{B}}$ volume element contributions and all $e=1,\cdots,n_{el}^{\partial\mathcal{B}}$ surface element contributions at $i=1,\cdots,n_{en}^{\mathcal{B}}$ respectively $i=1,\cdots,n_{en}^{\partial\mathcal{B}}$ element nodes to the global node points $I=1,\cdots,n_{np}$. Note, that q and $\overline{q}^{\mathrm{N}}$ are assumed to be temperature- independent throughout this work. With the definitions in Equation (17), Equation (16) represents the discretised nonlinear temperature field equation. The solution is performed by an incremental Newton-Raphson scheme, see [17] for detailed background information in the context of the finite element method. For that purpose, the Jacobian of the residual with respect to the temperature θ has to be determined. Therefore, the iterative scheme can be expressed as

$$r_{I,n+1}^{h,k+1} = r_{I,n+1}^{h,k} + \mathrm{d}r_I \quad \forall I = 1,\cdots,n_{np}, \tag{18}$$

wherein $\mathrm{d}r_I$ is approximated by the linear term of a Taylor series expansion,

$$\mathrm{d}r_I = \sum_{J=1}^{n_{np}} K_{IJ}\,\mathrm{d}\theta_J \quad \forall I = 1,\cdots,n_{np}. \tag{19}$$

The derivative $K_{IJ} = \partial r_I / \partial \theta_J$ is assembled to the global tangent operator

$$\boldsymbol{K}_{IJ}=\mathop{\mathbf{A}}_{e=1}^{n_{el}^{\mathcal{B}}}\int_{\mathcal{B}_{\mathcal{B}}^{e}}N_{X,\mathcal{B}}^{i}\left[c(\theta)\frac{1}{\Delta t}+\frac{\partial c(\theta)}{\partial\theta}\frac{\theta_{n+1}-\theta_{n}}{\Delta t}\right]N_{X,\mathcal{B}}^{j}\mathrm{d}V$$
$$+\mathop{\mathbf{A}}_{e=1}^{n_{el}^{\mathcal{B}}}\int_{\mathcal{B}_{\mathcal{B}}^{e}}\nabla_{X}N_{X,\mathcal{B}}^{i}\cdot\frac{\partial\boldsymbol{q}}{\partial\nabla_{X}\theta}\cdot\nabla_{X}N_{X,\mathcal{B}}^{j}\mathrm{d}V-\mathop{\mathbf{A}}_{e=1}^{n_{el}^{\partial\mathcal{B}}}\int_{\mathcal{B}_{\partial\mathcal{B}}^{e}}-N_{X,\mathcal{B}}^{i}\frac{\partial\overline{q}^{\mathrm{R}}(\theta)}{\partial\theta}N_{X,\mathcal{B}}^{j}\mathrm{d}A. \quad (20)$$

The first term represents the dynamic contribution and therefore characterises the time-dependency of the problem, whereas the second term corresponds to the internal heat flux where the constitutive relation for the heat flux $\boldsymbol{q}$ still has to be defined. The third term reflects the surface element contribution by Robin boundary conditions $\overline{q}^{\mathrm{R}}(\theta)$.

4. IMPLEMENTATION

Two aspects are vital for the computer aided planning of thermal spray processes: the prediction of the coating thickness distribution on the surface of a workpiece and the prediction of the temperatures reached in the workpiece during the process. For the simulation of the coating thickness, an efficient GPU-accelerated, C++ based simulation was developed in [6]. In the present work, this framework is modified so as to compute a time dependent temperature distribution θ^{∞} on the surface of the discretised workpiece mesh which serves as an input for the finite element framework developed in the previous section. In order to model the energy input of the unloaded flame of the spray gun into the workpiece, the surface heat flux is assumed to be applied by convection and radiation. An unloaded flame means that the problem at hand is restricted to heat transfer only without accounting for mass transport phenomena. One simple approach to mathematically cover the convective heat contribution is the introduction of a so-called film condition

$$\bar{q}_c = -h_c\left[\theta - \theta^\infty\right], \tag{21}$$

wherein h_c denotes the convective heat transfer coefficient which is assumed to be a constant, and where θ^∞ denotes the environmental temperature which either represents the HVOF spray gun or the room temperature. It should be noted that h_c represents a rather complex fluid mechanical process, not captured by the finite element approach as this work proceeds. The heat radiation is modelled by

$$\bar{q}_r = -\alpha_r\left[\theta - \theta^\infty\right] \quad \text{with } \alpha_r = \varepsilon\sigma\left[\theta^2 + \theta^{\infty 2}\right]\left[\theta - \theta^\infty\right], \tag{22}$$

where ε is the emissivity and σ represents the Stefan-Boltzmann constant, cf. Comini et al. [18] . Both, $\bar{q}_c$ and $\bar{q}_r$, are Robin boundary conditions in manner of Equation (4), i.e. $\bar{q}^{R}(\theta) = \bar{q}_c(\theta) + \bar{q}_r(\theta)$. For implementation, the heat capacity is specified by a cubic polynomial

$$c(\theta) = a_0 + a_1\theta + a_2\theta^2 + a_3\theta^3. \tag{23}$$

To guarantee thermal stability as discussed above, it has to be ensured that $c(\theta) > 0$ holds for arbitrary (positive) values of the absolute temperature θ. Furthermore, the heat conduction is modelled by Fourier's law

$$\boldsymbol{q} = -\boldsymbol{k}(\theta) \cdot \nabla_X \theta, \tag{24}$$

where $\boldsymbol{k}(\theta)$ has to be positive semi-definite for any θ, see e.g. [7] , and is thus represented by

$$\boldsymbol{k}(\theta) = \lambda(\theta)\boldsymbol{I} \quad \text{with } \lambda(\theta) = b_0 + \sum_{i=1}^{4} b_{i,1} \exp\left(-\left[\frac{\theta - b_{i,2}}{b_{i,3}}\right]^2\right), \tag{25}$$

where $\boldsymbol{I}$ characterises the second order identity tensor and where $\lambda(\theta) > 0$ is the heat conduction coefficient. The assumption of $\boldsymbol{k}(\theta)$ being proportional to $\boldsymbol{I}$ leads to an isotropic heat conduction model. The nonlinear heat conduction model specified via Equations (21)-(25) is implemented in C++ as part of the already mentioned simulation framework. The remaining part of this section describes the computation of the film condition parameters θ_i^∞ for all boundary triangles of the tetrahedron volume mesh representing the workpiece.

Apart from the triangulated workpiece mesh, the simulation approach takes a robot guided spray gun movement path as an input, which is represented as a sequence of 3-tuples $p_i = \{x_i, q_i, t_i\}$ for $i = 1, \cdots, n$ discrete robot path positions. Therein x_i is the spray gun position in global coordinates and q_i is a quaternion describing the orientation of the spray gun at time t_i. For detailed background information on quaternions the reader is referred to Argyris [19] , Betsch et al. [20] , Altmann [21] and the references cited therein.

During the simulation a virtual spray gun is moved along the given path in discrete time steps Δt, where positions are interpolated linearly and where quaternion slerp interpolation is used for the orientations [22]. For every gun position p_i the following computations, where steps 1-3 are entirely GPU-based and implemented in OpenGL and the OpenGL Shading Language (GLSL), are performed:

1) Compute the subset N_{loaded} of nodes of the workpiece mesh that receive a heat flux from the current spray gun position.

This is defined to be the subset of nodes, the coordinates of which are inside the spray plume, approximated by a (spray) cone which expands from the gun center position $x(t)$, see Figure 2. Only those nodes visible from the spray gun center are considered; nodes on the back side of the workpiece or nodes shadowed by other parts of the workpiece do not receive a high temperature load.

This visibility computation for arbitrary meshes with potentially hundreds of thousands of nodes and faces can be efficiently implemented by exploiting the computational power of modern graphics cards, tailored by design to perform visibility computations on large triangle meshes in real-time. For the rendering of virtual scenes consisting of triangle meshes, the graphics card projects the three dimensional nodes of the mesh in a perspective correct manner onto a two-dimensional image plane. In the first visibility computation step, projected nodes outside the rectangular screen area are discarded, limiting the remaining nodes to the ones inside a pyramidal viewing frustum as depicted in Figure 2. In an additional step, the graphics hardware uses the depth-buffer algorithm [23] to remove any nodes that are shadowed by the geometry for the particular viewpoint considered.

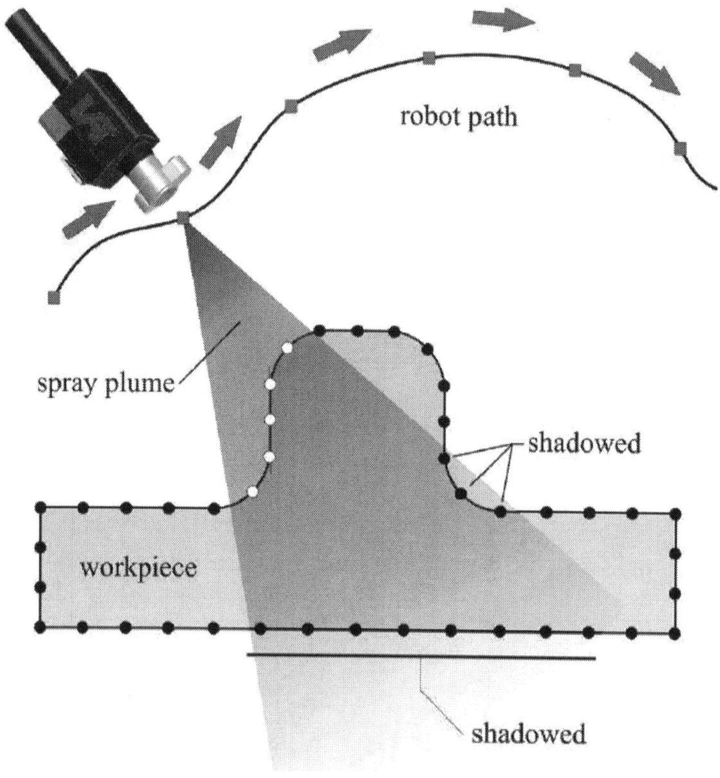

Figure 2. Visibility computation to determine the nodes of the workpiece to be coated and to receive a heat flux from the gun (white nodes). Nodes outside the spray cone or nodes that are shadowed from the gun by parts of the workpiece are not coated.

For the current spray gun position, first, the opening angle of the viewing frustum of the virtual camera is adjusted to match the desired opening angle of the spray cone. Subsequently, the workpiece mesh is rendered and the desired nodes that receive a heat flux from the gun are determined. The case that the viewing frustum is rectangular, but the spray cone should be circular is corrected in the next step.

2) For every node in the subset N_{loaded} determine the thermal load $\theta^{\infty}_{\text{node}}$ in order to set up Robin type boundary conditions.

The nodes are projected onto a plane orthogonal to the central axis of the spray cone at the stand off distance d_{SOD} from the gun center. The thermal load for every node is then computed based on its distance r_{no} from the cone center axis. Due to the lack of precise measurement data, the load function is modelled as a rotationally symmetric Gaussian with standard deviation s and amplitude $\theta_{\max}$ which is offset by r_{in} in order to cover a larger circular area of magnitude $\theta_{\max}$:

$$\theta^{\infty}_{\text{node}} = \begin{cases} \theta_{\max}, & \text{if } r_{\text{no}} < r_{\text{in}} \\ \theta_{\text{amb}}, & \text{if } r_{\max} \leq r_{\text{no}} \\ \theta_{\Delta\text{ma}} \exp\left(-\dfrac{1}{2}\left[\dfrac{r_{\text{no}} - r_{\text{in}}}{s}\right]^2\right) + \theta_{\text{amb}}, & \text{else} \end{cases} \tag{26}$$

with $\theta_{\Delta\text{ma}} = \theta_{\max} - \theta_{\text{amb}}$. The threshold $r_{\max}$ is used to limit the function to a circular radius thereby transforming the rectangular viewing frustum into the desired circular spray plume approximation. For the simulations presented in this article an HVOF spraying process is assumed and the values $d_{SOD} = 170\text{ mm}$, $\theta_{max} = \theta_{gun}^{\infty}$, $r_{in} = 10\text{ mm}$, $r_{max} = 66\text{ mm}$ and $s = 2\text{ mm}$ are used. The shape of the load function is shown in Figure 3.

3) The remaining nodes are set to receive an ambient thermal load θ_{amb}^{∞} in order to set up Robin type boundary conditions.

4) The thermodynamic model is evaluated.

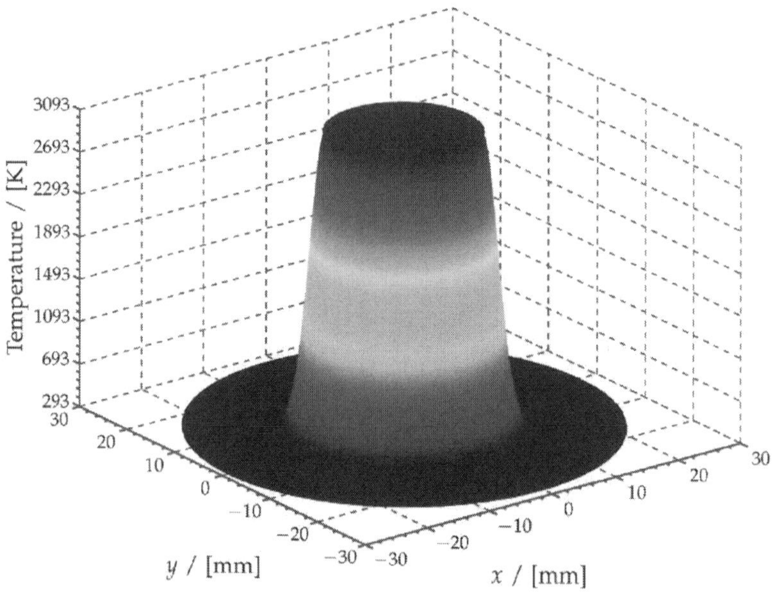

Figure 3. Shape of the radially symmetric load function θ_{node}^{∞} used in the benchmark simulations; Equation (26).

For each Newton iteration, the C++ implementation of the thermodynamic model assembles the global iterative residual vector $\boldsymbol{r}_I$ and the global tangent matrix $\boldsymbol{K}_{IJ}$ according to Equations (19) and (20), and then employs a conjugate gradient iterative solver with Jacobi (diagonal) preconditioning (pCG) to solve the system.

5) The simulation time index is increased by Δt.

This scheme is iterated until the end of the robot guided path is reached. A key aspect for the efficient computation of the boundary conditions is the similarity of the geometric relationships of the spray plume expanding from the gun nozzle towards the workpiece on the one hand, and the geometry-projection process towards a virtual camera center in the visibility computation on the other. Due to this similarity, the first three steps can easily be performed by exploiting GPU acceleration techniques and are implemented in OpenGL and the OpenGL Shading Language (GLSL) within our framework.

5. EXAMPLES

The implemented framework is now applied to carry out simulations of a realistic deep drawing tool in order to demonstrate the capabilities of the novel software. The starting temperature $\theta|_{t_0}$ of the material is considered to coincide with the room temperature $\theta^{\infty}_{\text{amb}} = 293\ \text{K}$ whereas the temperature of the spray gun is set to $\theta^{\infty}_{\text{gun}} = 3073\ \text{K}$. The steel material is St 35.8 (1.0305), the physical properties of which have been investigated by Richter [24] [25] up to a temperature of 873 K. The parameters for Equation (23), the polynomial of the heat capacity, are taken from [24] [25] and summarised in Table 1. Note, that the polynomial is parameterised in

°C not in K. For the sake of completeness, the polynomial is depicted with a K-scale in Figure 4(a). It can be seen that the parameter set is not valid for temperatures beyond the interval $\theta = [293;873]\,\mathrm{K}$. Since $c(\theta) > 0$ is guaranteed within the temperature range of interest, $\theta \geq 293\,\mathrm{K}$, the implemented framework remains thermally stable even when the physical valid temperature range is excelled. This turns out to be helpful for testing the algorithmic framework.

Furthermore, the parameters for the heat conduction coefficient $\lambda(\theta)$, given by Equation (25), are also summarised in Table 1. Because the parameters are the result of fitting a polynomial of Richter, [25] , both functions are depicted in Figure 4(b). The solid line represents the fitted function and, in contrast, the dashed line represents the polynomial taken from Richter, [25] . Obviously, the fit is very good within the physical valid temperature range. Beyond this range, the fitted curve asymptotically tends to a positive limit while the Richter polynomial has a root at $\theta = 1451.58\,\mathrm{K}$. As an advantage, the fitted function guarantees $\lambda(\theta) > 0$ for any θ, thereby avoiding numerical problems while testing the algorithmic framework. To complete the set of parameters, values for the convective heat transfer coefficients h_c^{air} and h_c^{gun} as well as for the emissivity ε and the Stefan-Boltzmann constant σ are summarised in Table 2. We assume the surface surrounded by air to be exposed to natural convection, as the surface loaded by the HVOF spray gun is assumed to be charged with forced convection.

The integrated coating and thermodynamic simulation is employed to predict the workpiece temperature for the manufacturing of thermal sprayed sheet metal forming tools. In this regard, the heat transport into a deep drawing tool of a moving and unloaded HVOF thermal spray gun is simulated using a complex robot tool path at three different movement speeds. The fastest simulation run completes the tool path within 40 s, the medium speed simulation in

80 s and the slowest run within 160 s. The tool path of the spray gun movement is depicted

Table 1. Parameters for $c(\theta)$ introduced in Equation (23) and $\lambda(\theta)$ introduced in Equation (25). The values are valid within $\theta = [293;873]\,\mathrm{K}$; see [25] for further details.

$c(\theta)$ a_0 / $[\mathrm{mJ \cdot g^{-1} \cdot K^{-1}}]$	a_1 / $[\mathrm{mJ \cdot g^{-1} \cdot K^{-2}}]$	a_2 / $[\mathrm{mJ \cdot g^{-1} \cdot K^{-3}}]$	a_3 / $[\mathrm{mJ \cdot g^{-1} \cdot K^{-4}}]$
4.22×10^{2}	9.31×10^{-1}	-2.14×10^{-3}	2.64×10^{-6}
$\lambda(\theta)$			
	b_0 / $[\mathrm{W \cdot mm^{-1} \cdot K^{-1}}]$		
	3×10^{-2}		
i	$b_{i,1}$ / $[\mathrm{W \cdot mm^{-1} \cdot K^{-1}}]$	$b_{i,2}$ / K	$b_{i,3}$ / K
1	1.549×10^{-2}	2.342×10^{2}	1.051×10^{2}
2	8.527×10^{-3}	3.717×10^{2}	1.357×10^{2}
3	2.091×10^{-4}	5.916×10^{2}	1.518×10^{1}
4	1.888×10^{-2}	5.389×10^{2}	3.266×10^{2}

Table 2. Values for the convective heat transfer coefficients h_c^{air} and h_c^{gun}, the emissivity ε and the Stefan-Boltzmann constant σ.

h_c^{air} / $[\mathrm{W \cdot mm^{-2} \cdot K^{-1}}]$	h_c^{gun} / $[\mathrm{W \cdot mm^{-2} \cdot K^{-1}}]$	ε	σ / $[\mathrm{W \cdot mm^{-2} \cdot K^{-4}}]$
1.0×10^{-5}	1.4×10^{-4}	0.81	5.670373×10^{-14}

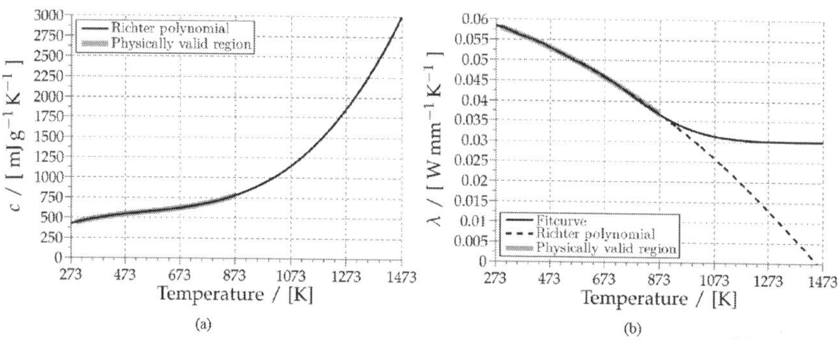

Figure 4. (a) Richter polynomial of the heat conduction $c(\theta)$; (b) Fitted curve and the Richter polynomial of the heat conduction coefficient $\lambda(\theta)$; see [25] for further details.

in Figure 5. The colors of the robot path represent the temperature load resulting from the respective gun position. For the robot path, the blue regions on the right and left parts have not been processed yet, since the gun started at the lower end of the path proceeding upwards along the circular opening of the workpiece onto the plain surface on top. For the path already processed, the coloring represents the surface load temperatures induced by the gun, where grey areas represent zero thermal load. In other words, these grey parts represent positions of the spray gun where the spray cone does not intersect with the workpiece. The surface colors represent the surface temperature at different time steps for the medium gun speed simulation run.

The simulation keeps track of the volume-averaged temperature and the peak workpiece temperature both with respect to time to identify critical sections of the path, which may cause an overheating of the workpiece. Figure 6 shows the simulated temperatures averaged over the entire workpiece volume and the peak temperature occurring in the workpiece—both plotted versus time—for the three gun speed simulation runs. The spraying process can be divided into four sections—compare Figures 5(b)-(d): First, the spray gun aims at the front radius working its way up to the plain surface, which is then coated in circular arc movements from left to right and back. During this process, peak temperatures are reached when the gun approaches the border of the workpiece and turns around for the next arc in opposite direction. The fluctuations of the peak temperature curves are due to the gun leaving and entering the workpiece at the borders. The third and fourth distinguishable sections are the right and left border areas of the workpiece including the screw holes for mounting the part. At the beginning of these sections, the peak and average temperatures increase considerably, because the large screw holes provide a significantly larger surface area that is heated by the spray gun. Afterwards the temperatures drop again when the gun gradually leaves the workpiece area. It can be seen that, for the considered robot tool path, the slowest simulated gun movement speed leads to the highest mean temperature of the workpiece as well as to the highest surface peak temperatures. The heat output to the environment through the surface is lower than the heat input by the spray gun.

6. SUMMARY

This work presents a novel integrated simulation approach for the computation of the workpiece temperature development during thermal spraying which is suited for large and arbitrarily complex shaped components. As its core, a thermodynamically consistent nonlinear heat transfer model based on the fundamental equations of continuum thermodynamics is employed. The equations obtained are transformed into their weak forms that represent the underlying equations of the presented non-linear finite-element-framework, implemented in C++ and the Eigen library. The material parameters are adjusted to experimental data by fitting suitable functions, and the parameters of the Robin boundary conditions are chosen in order to model realistic heat transfer. The presented simulations of a complex shaped deep drawing tool demonstrate a possible field of application for the novel software tool.

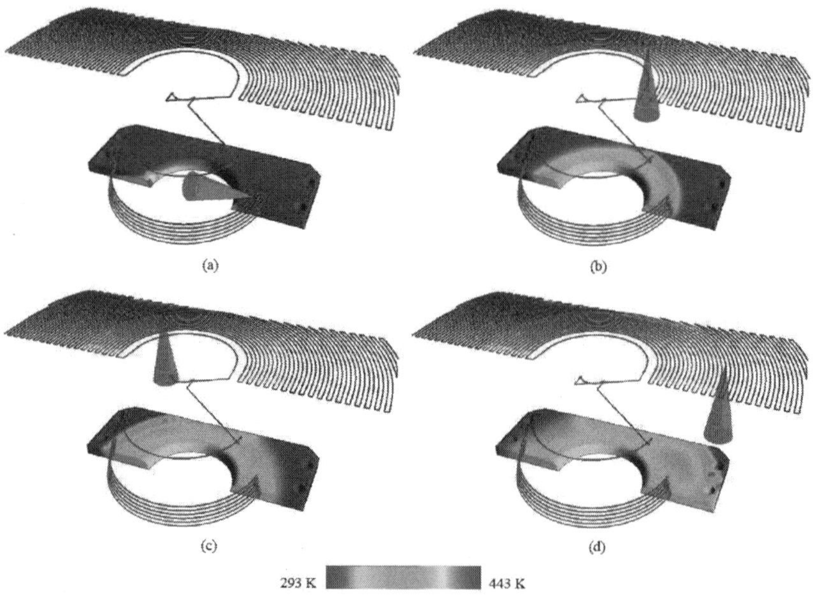

Figure 5. Simulated deep drawing workpiece and the employed robot movement path. The surface colors represent the surface temperatures and the path colors the temperature load after (a) 6 s; (b) 20 s; (c) 30 s; and (d) 40 s of the medium gun speed simulation run. Blue path segments have not been processed yet.

Future work will include the consideration of the mass flow of the HVOF gun in order to model the coating process of the substrate with hot particles. Furthermore, the developed framework shall be used to extend the optimisation software described in [26] in order to be able to optimise temperature quantities. Another aspect of future research is the extension towards a thermo-mechanically coupled framework, [27] , which will enable the analysis of residual stresses and damage due to thermal spraying as discussed in, e.g., [28] [29] .

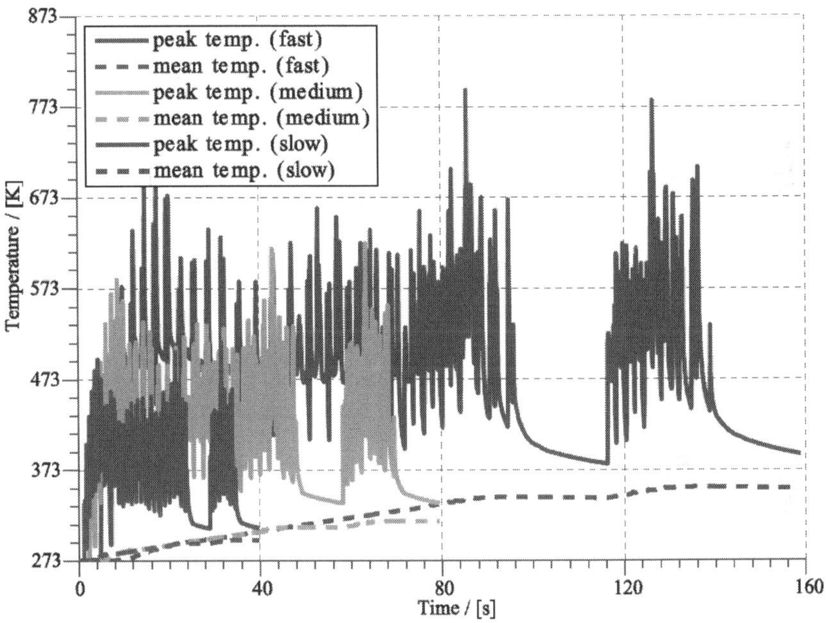

Figure 6. Simulated development of peak and mean workpiece temperature of the workpiece shown in Figure 5. The temperatures are depicted for the three gun speed simulation runs.

ACKNOWLEDGEMENTS

This research was funded by the German Research Foundation (DFG) as part of the collaborative research center 708 "3D-Surface Engineering of Tools for Sheet Metal Forming - Manufacturing, Modeling, Machining" within the projects B1 and B6.

REFERENCES

1. Trompeter, M., Franzen, V., Witulski, J. and Tekkaya, A.E. (2009) Thermisch beschichtete Werkzeuge für die Blechumformung. Der Schnitt- & Stanzwerkzeugbau, 5, 6-12.

2. Sahraoui, T., Fenineche, N.-E., Montavon, G. and Coddet, C. (2003) Structure and Wear Behaviour of HVOF Sprayed Cr3C2-NiCr and WC-Co Coatings. Materials & Design, 24, 309-313.

3. Thorpe, M. and H.J., R. (1992) A Pragmatic Analysis and Comparison of HVOF Processes. Journal of Thermal Spray Technology, 1, 161-170.

4. Fauchais, P., Vardelle, A. and Dussoubs, B. (2001) Quo Vadis Thermal Spraying? Journal of Thermal Spray Technology, 10, 44-66.

5. Coleman, B.D. and Noll, W. (1963) The Thermodynamics of Elastic Materials with Heat Conduction and Viscosity. Archive for Rational Mechanics and Analysis, 13, 167-178.

6. Wiederkehr, T., Müller, H., Krebs, B. and Abdulgader, M. (2009) A Deposition Model for Wire Arc Spraying and Its Computationally Efficient Simulation. Proceedings of the International Thermal Spray Conference (ITSC), Las Vegas, 492-498.

7. Liu, I.-S. (2002) Continuum Mechanics. Springer, Berlin.

8. Atefi, G. and Talaee, M.R. (2011) Non-Fourier Temperature Field in a Solid Homogeneous Finite Hollow Cylinder. Archive of Applied Mechanics, 81, 569-583.

9. Talaee, M.R. and Atefi, G. (2011) Non-Fourier Heat Conduction in a Finite Hollow Cylinder with Periodic Surface Heat Flux. Archive of Applied Mechanics, 81, 1793-1806.

10. Lewis, R.W., Nithiarasu, P. and Seetharamu, K.N. (2004) Fundamentals of the Finite Element Method for Heat and Fluid Flow. Wiley, Hoboken and N.J.

11. Bergheau, J.M. and Fortunier, R. (2008) Finite Element Simulation of Heat Transfer. Wiley-ISTE, London.

12. Kuhl, E., Denzer, R., Barth, F.J. and Steinmann, P. (2004) Application of the Material Force Method to Thermo-Hyperelasticity. Computer Methods in Applied Mechanics and Engineering, 193, 3303-3325.

13. Vujicic, M.R. (2006) Finite Element Solution of Transient Heat Conduction Using Iterative Solvers. Engineering Computations, 23, 408-431.

14. Palani, G. and Kim, K.Y. (2010) Numerical Study on a Vertical Plate with Variable Viscosity and Thermal Conductivity. Archive of Applied Mechanics, 80, 711-725.

15. Gross, M. and Betsch, P. (2011) Galerkin-Based Energy-Momentum Consistent Time-Stepping Algorithms for Classical Nonlinear Thermo-Elastodynamics. Mathematics and Computers in Simulation, 82, 718-770.

16. Bonet, J. and Wood, R. (2008) Nonlinear Continuum Mechanics for Finite Element Analysis. 2nd Edition, Cambridge University Press, Cambridge.

17. De Borst, R., Crisfield, M., Remmers, R. and Verhoosel, C. (2012) Nonlinear Finite Element Analysis of Solids and Structures (Wiley Series in Computational Mechanics). Wiley, Hoboken.

18. Comini, G., Del Guidice, S., Lewis, R.W. and Zienkiewicz, O.C. (1974) Finite Element Solution of Non-Linear Heat Conduction Problems with Special Reference to Phase Change. International Journal for Numerical Methods in Engineering, 8, 613-624.

19. Argyris, J. (1982) An Excursion into Large Rotations. Computer Methods in Applied Mechanics and Engineering, 32, 85-155.

20. Betsch, P., Menzel, A. and Stein, E. (1998) On the Parametrization of Finite Rotations in Computational Mechanics: A Classification of Concepts with Application to Smooth Shells. Computer Methods in Applied Mechanics and Engineering, 155, 273-305.

21. Altmann, S.L. (2005) Rotations, Quaternions, and Double Groups. Dover Books on Mathematics. Dover Publications, Mineola.

22. Shoemake, K. (1985) Animating Rotation with Quaternion Curves. SIGGRAPH Computer Graphics, 19, 245-254.

23. Catmull, E.E. (1974) A Subdivision Algorithm for Computer Display of Curved Surfaces. Ph.D. Dissertation, The University of Utah, Salt Lake City.

24. Richter, F. (1973) Die wichtigsten physikalischen Eigenschaften von 52 Eisenwerkstoffen: Mitteilung aus dem Forschungsinstitut der Mannesmann AG. Verl. Stahleisen, Düsseldorf.

25. Richter, F. (1983) Physikalische Eigenschaften von Stählen und ihre Temperaturabhängigkeit: Polynome und graphische Darstellungen. Stahleisen, Düsseldorf.

26. Kout, A. and Müller, H. (2009) Parameter Optimization for Spray Coating. Advances in Engineering Software, 40, 1078-1086.

27. Baek, S. and Srinivasa, A.R. (2003) Thermomechanical Constraints and Constitutive Formulations in Thermoelasticity. Mathematical Problems in Engineering, 2003, 153-171.

28. Boehmer, J., Funk, G., Jordan, M. and Fett, F. (1998) Strategies for Coupled Analysis of Thermal Strain History during Continuous Solidification Processes. Advances in Engineering Software, 29, 679-697.

29. Fagerstrom, M. and Larsson, R. (2008) A Thermo-Mechanical Cohesive Zone Formulation for Ductile Fracture. Journal of the Mechanics and Physics of Solids, 56, 3037-3058.

CHAPTER 2

Ductile Fracture Characterization for Medium Carbon Steel Using Continuum Damage Mechanics

Stergios Pericles Tsiloufas*, Ronald Lesley Plaut

Department of Metallurgical and Materials Engineering,
Escola Politécnica, University of São Paulo, São Paulo, Brazil.

ABSTRACT

This paper presents the ductility characterization for a medium carbon steel, for two microstructural conditions, that has been evaluated using the continuum damage mechanics theory, as proposed by Kachanov and developed by Lemaitre. Tensile tests were carried out using loading-unloading cycles in order to capture the gradual deterioration of the elastic modulus, which may be linked to the ductile damage increase with increasing plastic strain. The mechanical parameters for the isotropic damage evolution equation were obtained and then used as inputs for a plasticity-damage coupled numerical algorithm, validated through numerical simulations of the experimental tensile tests. A comparison between the SAE 1050 steels studied and two carbon steel alloys (obtained from the literature), provided some basic understanding of the influence of the carbon level on the evolution of the damage parameters. An empiric relationship for this set of parameters, which can provide useful data for preliminary studies envisaging prediction of ductile failure in carbon steels, is also presented.

KEYWORDS

Continuum Damage Mechanics; Tensile Testing; Numerical Simulation; Medium Carbon Steels

1. INTRODUCTION

Ductile fracture is the failure of a solid material due to nucleation, coalescence and growth of cavities induced by plastic deformation. There are several ways, from empiric relationships [1,2] to porous media stiffness modeling [3], developed to study this phenomenon in order to predict when a workpiece will fail under a given stressstrain state.

Kachanov, in 1958 [4], first proposed a continuum damage variable to represent the surface density of cavities in a given infinitesimal volume element. By the 70's, researchers embraced the idea and developed a theory based on the framework of irreversible processes thermodynamics to model the evolution of this damage variable and how it affects mechanical properties, such as elastic modulus and stresses, leading to the eventual failure of a material [5]. This theory, called Continuum Damage Mechanics (CDM), is complementary to Fracture Mechanics, since it is concerned about the nucleation and growth of cavities until they reach a critical size turning into a macroscopic crack, whose propagation in a solid media is studied by the latter.

For damage evolution caused by large plastic deformation, Lemaitre and Chaboche developed the first and simplest model [6-10], which considers a linear evolution of isotropic damage with plastic strain in a uniaxial stress state condition. This model was later expanded by other authors, adding new capabilities such as dealing with anisotropic damage [11], or with non-linear damage evolution [12-16].

Recently, medium carbon steel heat-treated to obtain spheroidized cementite in its microstructure is being used as raw material for sheet forming processes, due to its better formability properties [17]. Due to large plastic strains imposed to this kind of manufactured parts, cracks and other defects observed are mostly related to ductile damage evolution.

With this motivation, in this work the ductile fracture of SAE 1050 steel was studied, for two different microstructural conditions namely: lamellar ferrite-pearlite and spheroidized cementite, under the continuum damage mechanics point of view. Experimental characterization of isotropic damage evolution was carried out and numerical simulations were performed in order to predict failure.

2. CONTINUUM DAMAGE MECHANICS MODEL FOR DUCTILE FRACTURE

The continuum damage variable introduced by Kachanov is defined as the relationship between the sectional area of voids A_v and the overall sectional area A_0 of a given surface in a volume element.

Assuming the hypothesis of isotropic deterioration of the material, the damage can be written as:

$$D = \frac{A_v}{A_0} = 1 - \frac{A_{eff}}{A_0} \qquad (1)$$

where A_{eff} is the effective resisting area. It may be observed that this definition is the same one as the microvoid area fraction used is some micromechanical theories, such as McClintock's [18]. The damage variable can assume any value between 0 and 1, covering from a virgin state to a completely damaged one, although real materials will fail when the damage reaches a critical value $D_c < 1$, when the effective area can no longer resist the applied load, leading to the formation of a macroscopic crack.

In his model, Lemaitre assumes the hypothesis of strain equivalence, which states that the damaged material will have the same constitutive behavior of the virgin material, replacing the stress tensor σ by the effective stress tensor $\tilde{\sigma}$, defined as:

$$\tilde{\sigma} = \frac{\sigma}{1 - D} \qquad (2)$$

One important consequence of this assumption is that one can define an effective elastic modulus of a damaged material, giving an indirect way to measure the damage in a solid, by monitoring the evolution of the Young modulus with increasing strain:

$$D = 1 - \frac{\tilde{E}}{E} \tag{3}$$

where $\tilde{E}$ is the effective elastic modulus and E is the elastic modulus for the undamaged material.

Using as basis the thermodynamic of irreversible processes [19], CDM treats the damage as an internal thermodynamic state variable, and so its evolution can be derived assuming the existence of a potential of dissipation ϕ and an associated variable Y, named damage strain energy release rate and defined as [10]:

$$-Y = \frac{\sigma_{eq}^2}{2E(1-D)^2}\left[\frac{2}{3}(1+\nu)+3(1-2\nu)\left(\frac{\sigma_H}{\sigma_{eq}}\right)^2\right] \tag{4}$$

where $\sigma_{eq} = \sqrt{2/3}\left\|\sigma^D\right\|$ is the von Mises equivalent stress, σ^D is the deviatoric stress tensor, ν is the Poisson's ratio and $\sigma_H = (1/3)tr(\sigma)$ is the hydrostatic stress. Further, Lemaitre [10] shows that the damage evolution can be written as:

$$\dot{D} = -\frac{\partial \phi}{\partial Y}(1-D)\dot{p} \tag{5}$$

with $p=\sqrt{2/3}\|\varepsilon^{p}\|$ defined as the accumulated plastic strain and ε^{p} being the plastic strain tensor. The choice of a proper potential of dissipation that can represent experimental results is the core of any CDM model. In Lemaitre and Chaboche's model, the hypothesis of isotropic damage, existence of a strain threshold for damage initiation and linear evolution of the damage with the accumulated plastic strain leads to the following equation for damage evolution:

$$\dot{D}=\begin{cases} 0 & \text{for } p<p_D \\ \dfrac{-Y}{S}\dot{p} & \text{for } p\geq p_D \end{cases} \qquad (6)$$

where p_D is the accumulated plastic strain threshold and S is the damage resistance parameter, which are material dependent properties. For the uniaxial stress state, and assuming that the elastic strain can be neglected in comparison to the total strain, the accumulated plastic strain can be considered equal to the principal strain. The damage increases until it reaches a critical value D_c which can be calculated with the following equation [10]:

$$D_c=\frac{D_{1c}}{R_v}\left[\frac{\sigma_u}{\sigma_{eq}}(1-D)^2\right] \qquad (7)$$

where D_{1c} is the critical damage for the uniaxial stress state and can be measured in a tensile test, σ_u is the ultimate tensile stress and

$$R_v = \left[(2/3)(1+\nu)+3(1-2\nu)(\sigma_H/\sigma_{eq})^2\right]$$

is called triaxiality factor, which accounts for the difference between the actual stress state and the perfectly uniaxial stress state.

This model was later implemented in the Abaqus/Explicit solver using the VUMAT subroutine [20] following the numerical algorithm proposed by Lee and Pourboghrat [21].

3. EXPERIMENTAL PROCEDURE

In order to determine mechanical properties and damage parameters, standard tensile tests were carried out for specimens of SAE 1050 steel for the lamellar and for the spheroidized microstructures. Three specimens for the hot rolled ferrite-pearlite material and nine specimens for the spheroidized material were tested.

The spheroidized material has been cold rolled, with a thickness reduction of 50% and subsequently annealed at 700°C for 13 hours in a 100% H_2 atmosphere, to obtain the characteristic spheroidized microstructure. The specimens were machined from a 1.0 mm thickness sheet (spheroidized material) and from a 2.0 mm thickness sheet (hot rolled material. The neck section had 75 mm length and 12.5 mm width, as shown in Figure 1.

The damage variables were calculated using the variation of the elastic modulus, so several loading-unloading cycles were needed in order to measure this property with strain increase. The tests were performed in an Instron 3369 universal testing machine, with a 50 kN load cell. Each cycle began with a 1 mm crosshead displacement followed by an unloading until the force attained 50 N. The crosshead velocity was fixed at 2 mm/min. The strains were measured through a clip gage extensometer with 50 mm gage length. Sampling frequency was 5 Hz. Figure 2 illustrates the experimental setup.

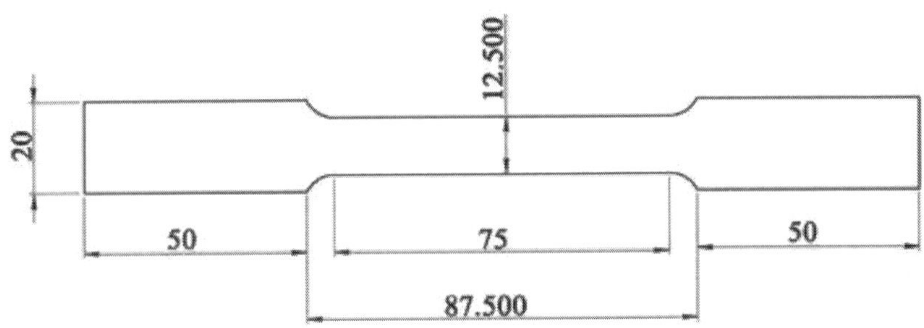

Figure 1. Workpiece dimensions, in mm, for the tensile test.

Figure 3 shows the true stress-strain curves for both types of tested specimens. The loading-unloading cycles shown were used in the evaluation of the elastic modulus, measured always during the unloading path, following recommendations by Lemaitre [10]. The drop in the true stress, as pictured in this figure, can be linked to the fracture initiation. To represent the work hardening behavior of the material, Ludwik equation [22] has been used, as presented by Equation (8). The material constants are given in Table 1.

$$\sigma_Y\left(\varepsilon^p\right)=\sigma_0+K\left(\varepsilon^p\right)^n \tag{8}$$

The evolution of the elastic modulus is shown in Figure 4 for the tested materials (including the pure iron results [23] and other carbon steels [24,25] obtained from the literature). It may be observed that the elastic modulus decreases with increasing carbon level. Also, there is a significant non-linear drop in the elastic modulus for small strains, followed by a linear evolution. Lemaitre's model considers that the damage does not occur for a strain below the critical value, and will grow with a constant rate after that value. For this reason, following the same procedure of Celentano et al. [24], any elastic modulus degradation below the linear part of the curve will be neglected. This transition coincides with the transition of the elastic regime to the plastic behavior of the material. Therefore, yielding strain will be considered as the damage strain threshold and the elastic modulus, at this point, will be assumed to be the one for the undamaged material.

The damage evolution, measured trough Equation (3), is shown in Figure 5 for both studied alloys. Critical damage D_{1c} is taken as the damage value prior to the non-linear increase in damage, just before fracture. The other parameter to be evaluated is the damage resistance S, calculated trough Equation (9), which is obtained by manipulating Equation (6) and assuming that in the tensile test the material is under a perfectly uniaxial stress state.

$$S = \frac{\sigma_{eq}^2}{2E(1-D)^2(dD/d\varepsilon)} \tag{9}$$

To obtain the value of S, several experimental points must be taken from Figures 3 and 5 for different strains.

Table 1 summarizes the mechanical and damage parameters identified for SAE 1050 steel for both microstructural conditions.

It must be pointed out that these parameters can be used as inputs for finite element simulations of the tensile test.

(a) (b)

Figure 2. Experimental setup for the tensile tests.

4. NUMERICAL SIMULATIONS

Lemaitre's model was implemented in Abaqus/Explicit finite element solver using a VUMAT subroutine aiming at the coupling of isotropic plasticity with damage, based on the stress integration algorithm called operator-split.

For each time step, the incremental strain was considered as being fully elastic, and then the corresponding stress tensor was evaluated. The von Mises criterion, coupled with damage, was used to determine if the material is indeed below the yielding condition:

$$\frac{\sigma_{eq}}{1-D} \leq \sigma_Y\left(\varepsilon^p\right) \tag{9}$$

If Equation (10) is not satisfied, then a plastic correcting procedure must be used to calculate the plastic increment and ensure the consistency condition. Details of this plastic corrector can be found in Lee and Pourboghrat [22]. After this calculation, stresses, the damage variable and the plastic strain are updated for the next step. Further details may be obtained in Tsiloufas [26].

The tensile test was simulated using an imposed longitudinal displacement on the right end of the specimen, with the same 2 mm/min velocity as for the experimental procedure. Boundary conditions of restricted transversal and normal displacement were imposed for both ends of the specimen, simulating the jaws of the tensile test equipment. The mesh in the test region is formed by 4500 solid hexahedral elements, with 8 nodes, linear integration and 0.5 mm length.

Figure 6 shows the resulting true stress-strain curves. The numerical simulations could properly reproduce the work hardening behavior of the tested materials, however the difference between the effective stress $\tilde{\sigma}_{eq} = \sigma_{eq}/(1-D)$, that is measured through the load cell in the experiments, and the real stress σ_{eq}, must be pointed out. This may be explained by observing the material work hardening curve in terms of the real stress reaching saturation, due to the damage increase, which diminishes the material's area of resistance.

In more detail, damage evolution is shown in Figure 7. It may be observed that damage does not evolve in a linear way, as expected by Lemaitre's theory. This can be explained observing Figure 8, which shows the evolution of the triaxiality factor R_v. Lemaitre's model considers that the stress state in a tensile test is perfectly uniaxial, which is not true, as the value of R_v grows for strains higher than 0.15.

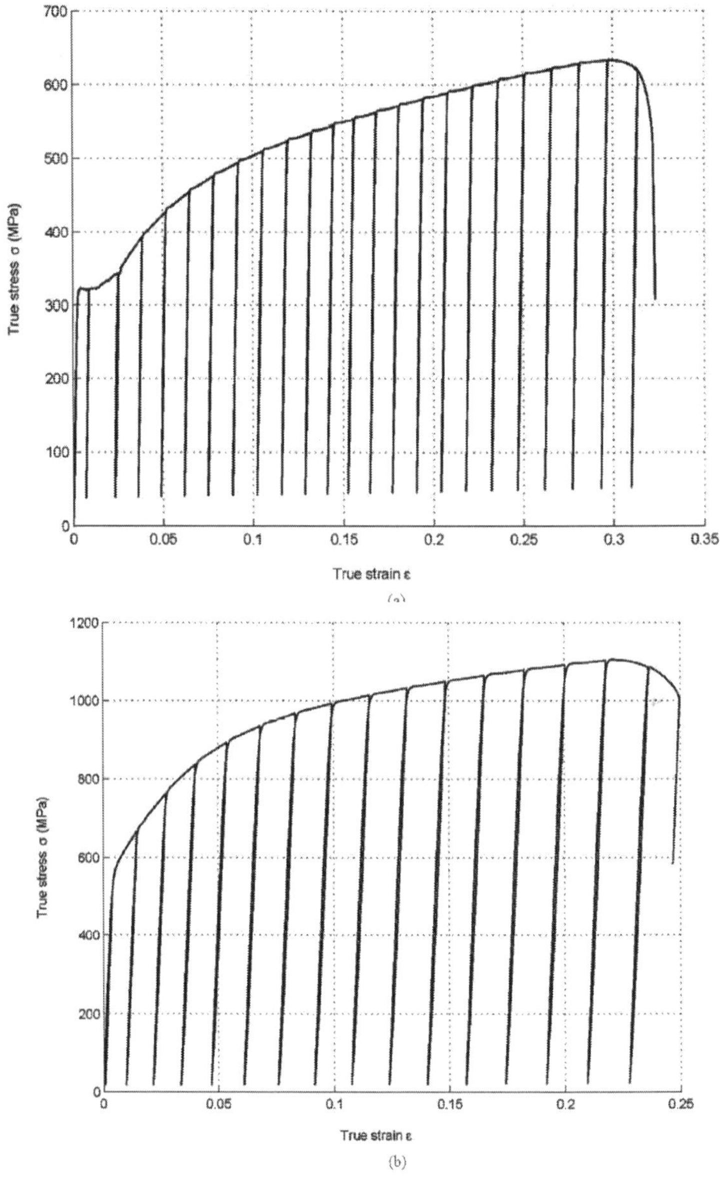

Figure 3. True stress-strain curves for SAE 1050 steel. (a) Spheroidized alloy, (b) Hot rolled alloy.

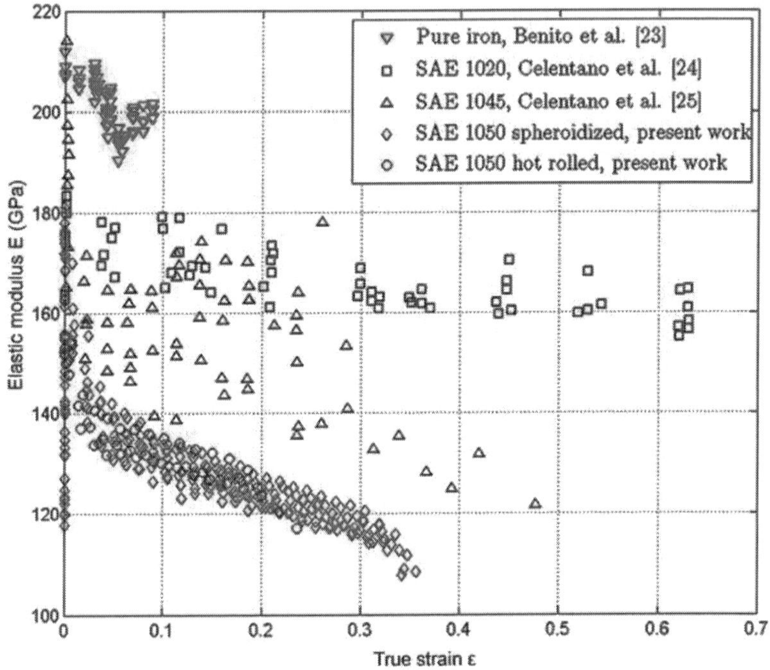

Figure 4. Elastic modulus evolution for SAE 1050 steel and comparison with pure iron and other carbon steels [23-25].

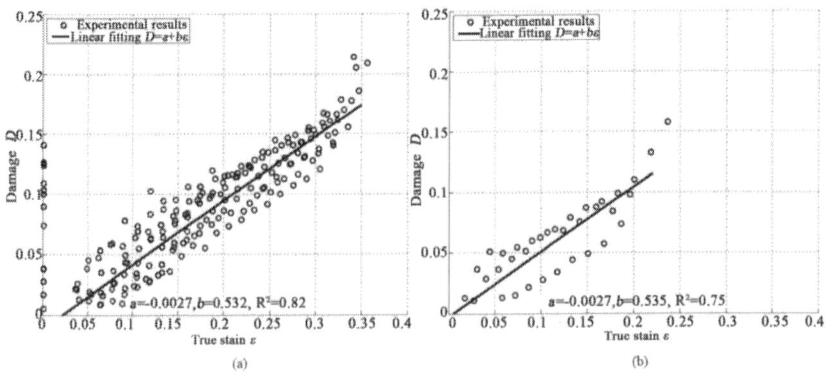

Figure 5. Ductile damage evolution versus true strain for SAE 1050. (a) Spheroidized alloy, (b) Hot rolled alloy.

Table 1. Mechanical and damage parameters for SAE 1050 steel

	Spheroidized	Hot rolled
E (GPa)	138	139
V (-)	0.29	1.29
σ_0 (MPa)	290	506
σ_u (MPa)	595	1073
K (MPa)	520	998
N (-)	0.41	0.33
S (MPa)	1.97	4.79
ε_D (-)	0.022	0.005
D_{1c} (-)	0.19	0.13

Figures 9 and 10 shows the damage contour evolution for both alloys, along with a picture of the fractured experimental specimen. The fracture in the experimental test does not occur in the middle part of the sample, because any imperfection in the manufacturing of the workpiece or in the setup of the experiment may change the fracture position. It may be observed that damage is localized in the region near to the fracture, which is expected, since the localization of the strain occurring in the necking zone is the main responsible for the damage accumulation.

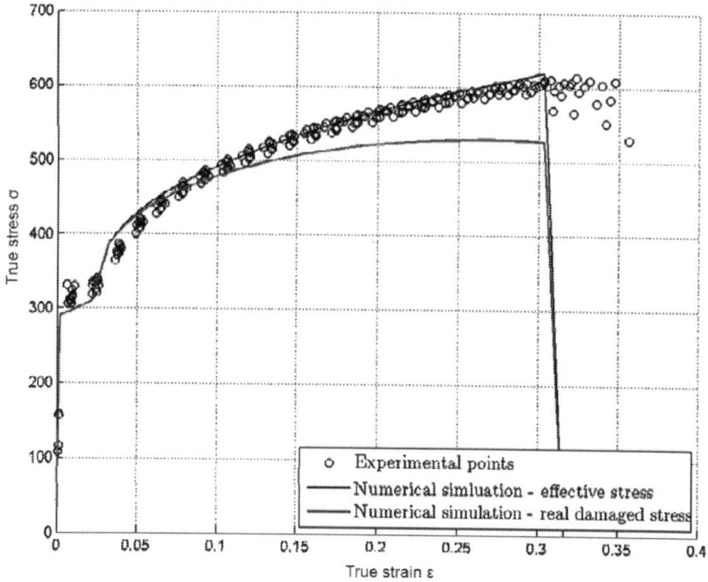

(a)

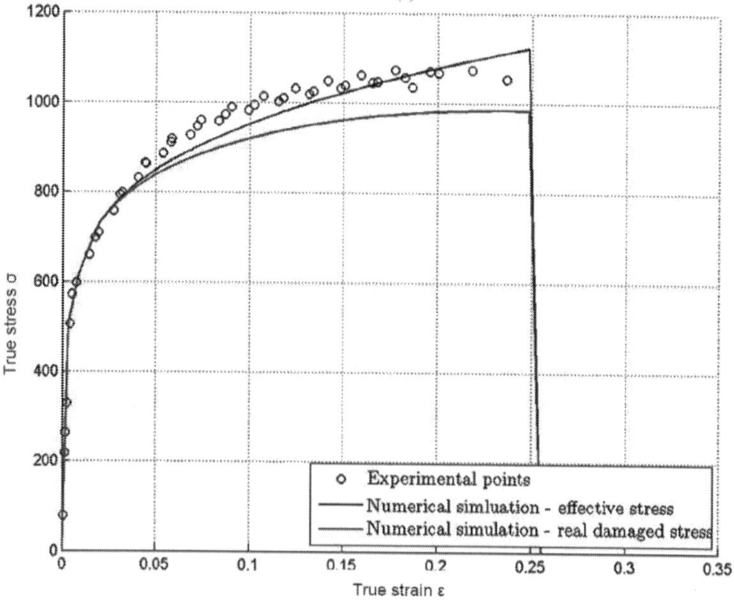

(b)

Figure 6. Strain hardening curves for SAE 1050. Experimental and numerical results. (a) Spheroidized alloy, (b) Hot rolled alloy.

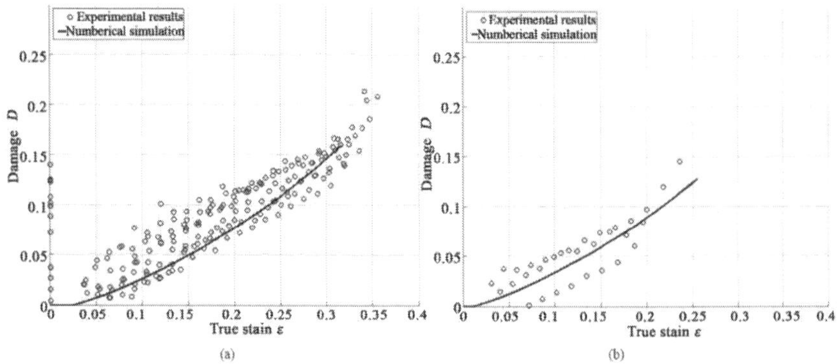

Figure 7. Damage evolution with strain for SAE 1050. Experimental and numerical results. (a) Spheroidized alloy, (b) Hot rolled alloy.

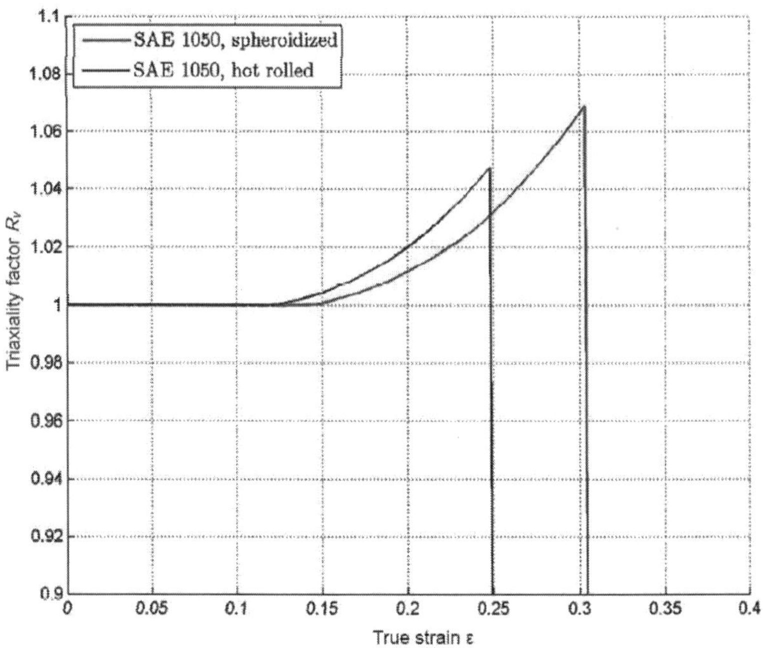

Figure 8. Evolution of the triaxiality factor for both tested materials.

5. DAMAGE BEHAVIOR FOR CARBON STEEL ALLOYS

Using data from Refs. [24,25], Figure 11 summarizes the damage evolution in a tensile test for SAE 1020 and for 1045 steels and those of the present research. It is possible to notice that the strain threshold for damage initiation decreases and the damage growth increases with increasing carbon levels. The reason behind this fact can be associated to the microvoid formation in ductile fracture, which nucleates and grows at the interface between the ferrite matrix and the harder second phase particles [27], whose amount is proportional to the carbon level. Also, the microstructure seems to act mostly on the value of the critical damage, since for both tested SAE 1050 steels, damage initiates and increases in a similar fashion, although they present different critical damage values.

For practical/industrial purposes, it is interesting to suggest a curve fitting for both damage resistance S and the damage strain threshold ε_D, as a function of the carbon level for SAE 10XX steels. Results are shown in Figure 12. A power law of the form $y = ax^b + c$, with constants given in Table 2, has been chosen because of good agreement with experimental points.

These curves can be used as guidelines for numerical simulations in industry as a first approximation in a basic project. The uniaxial critical damage D_{1c} can be calculated using the empiric relationship of Lemaitre [10]:

$$D_{1c} = 1 - \frac{s_f}{s_u} \qquad (10)$$

where s_f is the engineering fracture stress and s_u is the ultimate tensile engineering stress, both parameters easily provided by steel manufacturers. A comparison between the estimated and the measured value for uniaxial critical damage is shown in Table 3.

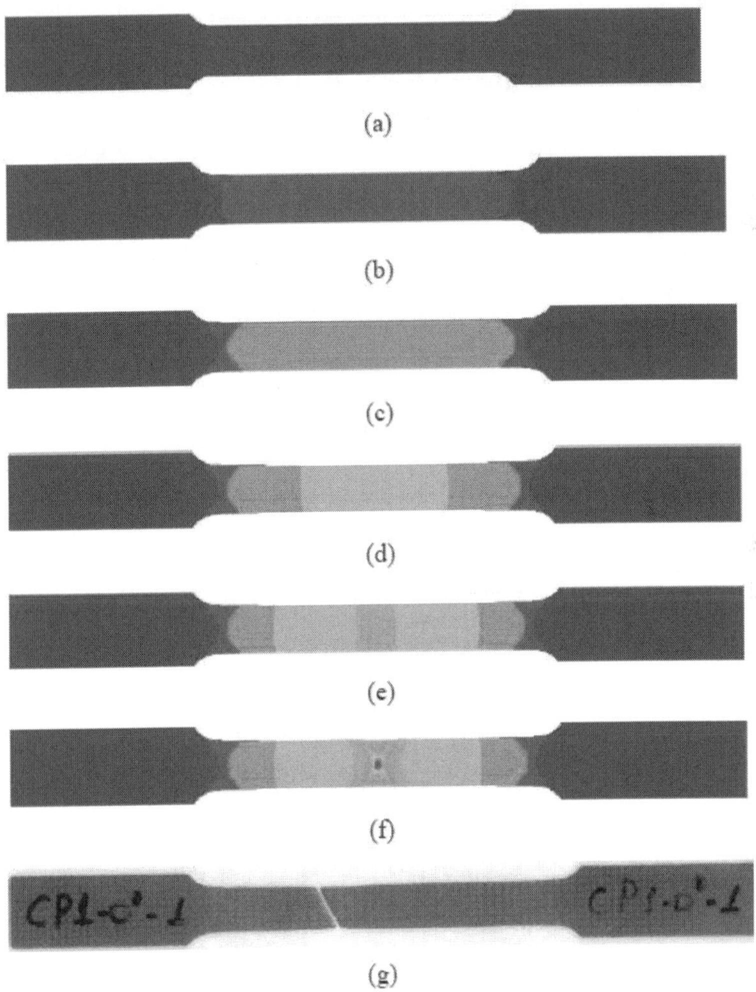

Figure 9. (a)-(f) Damage contour evolution (g) fractured experimental specimen. Spheroidized SAE 1050 steel.

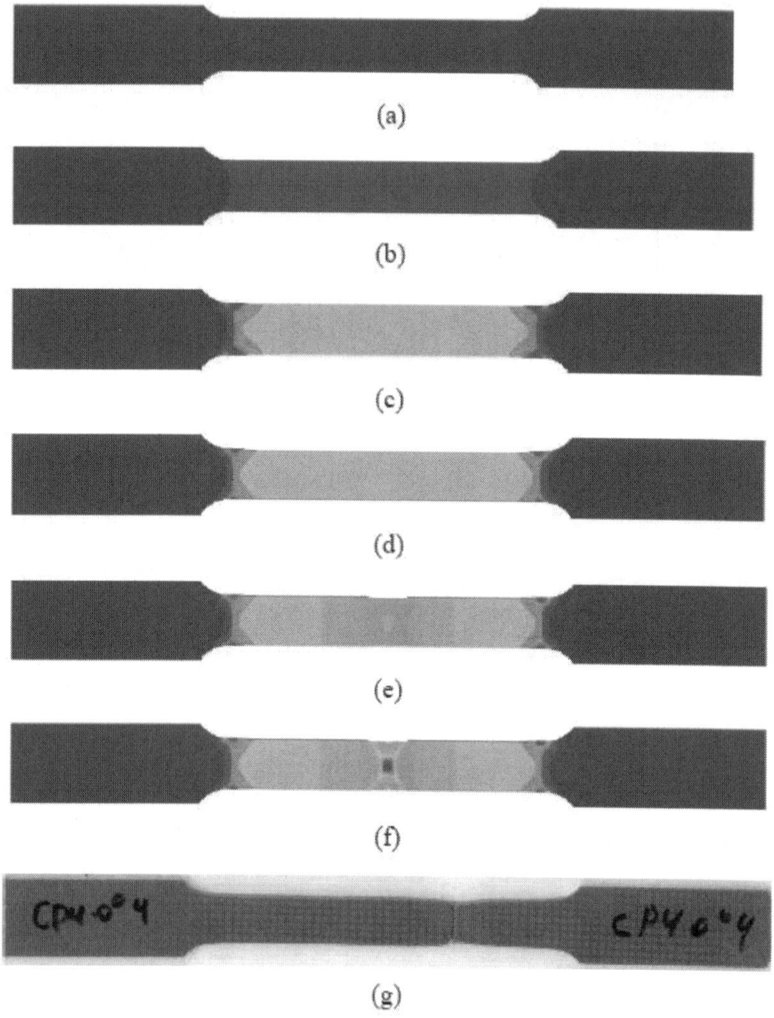

Figure 10. (a)-(f) Damage contour evolution (g) fractured experimental specimen. Hot rolled SAE 1050 steel.

Although these guidelines can be useful, a parameter characterization procedure similar to the one here presented is highly recommended, for a more precise result.

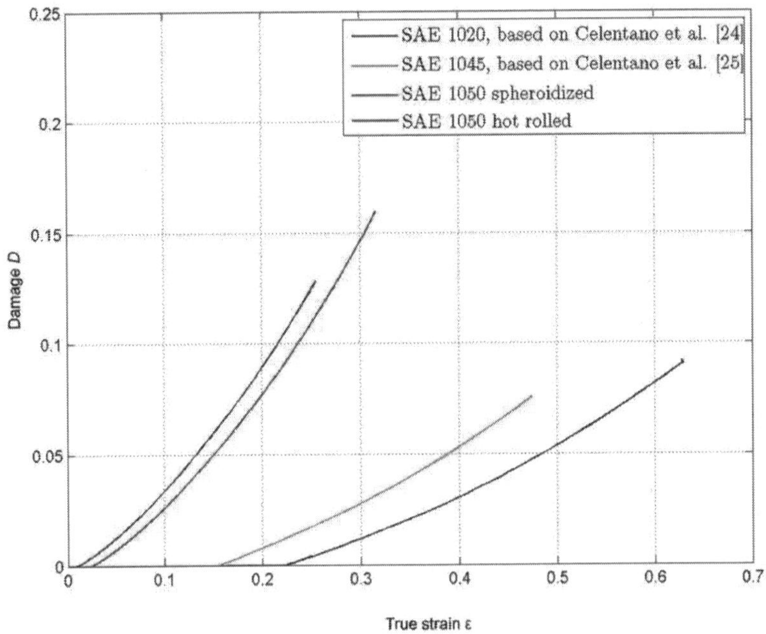

Figure 11. Damage evolution for carbon steel alloys.

6. CONCLUSIONS

In the present work, a continuum damage characterization study using loading-unloading cycles during tensile testing was performed for SAE 1050 steel for two microstructural conditions: heat-treated spheroidized cementite and lamellar ferrite-pearlite (hot rolled). Damage itself was indirectly measured by means of the degradation of the elastic modulus. Mechanical parameters for modeling the work hardening, behavior and the damage evolution were evaluated for the Lemaitre's ductile damage model.

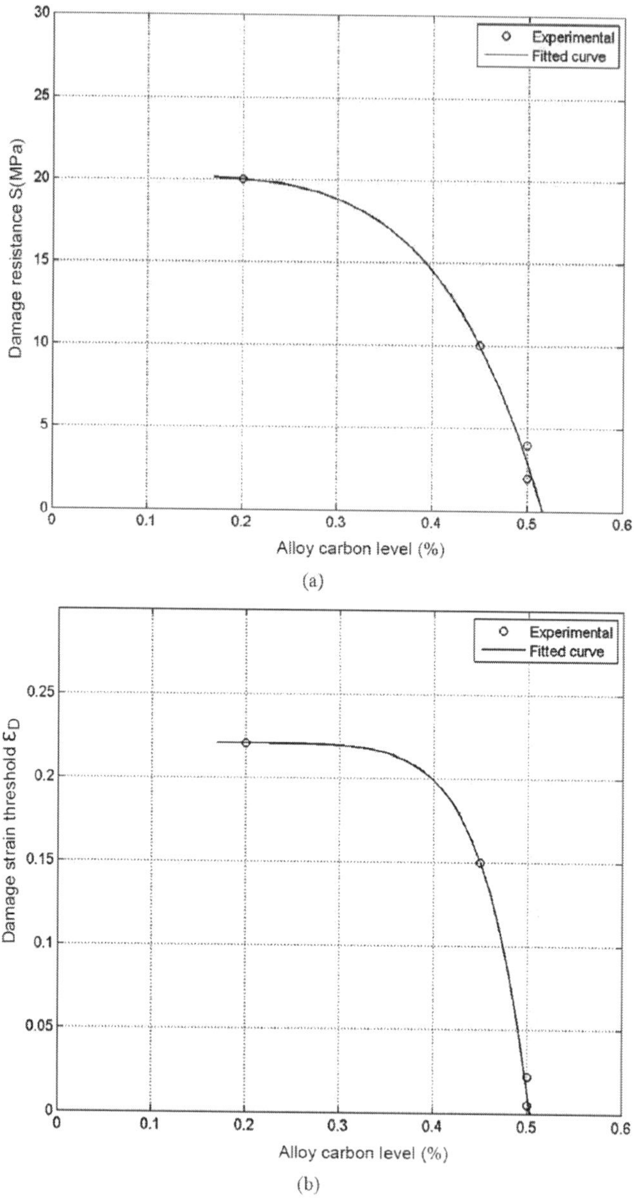

Figure 12. Power law fitting for damage properties as a function of the carbon level in SAE 10XX steels. (a) Damage resistance, (b) Damage strain threshold.

Table 2. Power law constants for damage parameters as a function of carbon level for SAE 10XX steels.

Property	a	b	c
Damage resistance S (MPa)	−537.2	4.97	20.2
Damage strain threshold ε_D	−252.5	10.3	0.220

Table 3. Measured and estimated uniaxial critical damage for SAE 1050 steel.

	Spheroidized	Hot rolled
Ultimate engineering stress s_u (MPa)	480	790
Fracture engineering stress s_f (MPa)	400	900
Estimated critical damage D_{1c}	0.17	0.12
Calculated critical damage D_{1c}	0.19	0.13
Error in critical damage estimation	10.5%	7.7%

A numerical algorithm was written to account for the coupling between damage and isotropic plasticity, and implemented in Abaqus/Explicit solver by means of a VUMAT subroutine. Then, simulations of the tensile test were performed, providing good agreement with experimental results. The difference between effective stresses and the real stresses acting on a damaged volume element is presented. Numerical damage evolution was not linear with strain, as would be expected. The explanation lies on the stress-state, which ceases to be perfectly uniaxial for a strain higher than about 0.15.

Although damage mechanics cannot describe macrocrack formation that takes place just prior to fracture, using such a model together with experimental characterization procedure seem to be a useful manner for predicting the initial stages of a ductile failure phenomenon.

ACKNOWLEDGEMENTS

The authors would like to thanks Brasmetal Waeholtz for providing the material samples for the tests, GMSIEPOLI/USP for the tensile test machine and CAPES for the scholarship of SPT, provided for the development of this study.

REFERENCES

1. M. G. Cockroft and D. J. Latham, "Ductility and the Workability of Metals," Journal of the Institute of Metals, Vol. 96, 1968, pp. 33-39.

2. M. Oyane, T. Sato, K. Okimoto and S. Shima, "Criteria for Ductile Fracture and Their Applications," Journal of Mechanical Working Technology, Vol. 4, No. 1, 1980, pp. 65-81.

3. A. L. Gurson, "Continuum Theory of Ductile Rupture by Void Nucleation and Growth: Part I—Yield Criteria and Flow Rules for Porous Ductile Media," Journal of Engineering Materials and Technology, Vol. 99, No. 1, 1977, pp. 2-15.

4. L. M. Kachanov, "Rupture Time under Creep Conditions," International Journal of Fracture, Vol. 97, 1999, pp. 11-18.

5. J. L. Chaboche, "Continuum Damage Mechanics: Present State and Future Trends," Nuclear Engineering and Design, Vol. 105, No. 1, 1987, pp. 19-33.

6. J. Lemaitre, "How to Use Damage Mechanics," Nuclear Engineering and Design, Vol. 80, No. 1, 1984, pp. 233- 245.

7. J. Lemaitre, "A Continuous Damage Mechanics Model for Ductile Fracture," Journal of Engineering Materials and Technology, Vol. 77, 1985, pp. 335-344.

8. J. L. Chaboche, "Continuum Damage Mechanics: Part I— General Concepts," Journal of Applied Mechanics, Vol. 55, No. 1, 1988, pp. 55-64.

9. J. L. Chaboche, "Continuum Damage Mechanics: Part II— Damage Growth, Crack Initiation and Crack Growth," Journal of Applied Mechanics, Vol. 55, No. 1, 1988, pp. 65-72.

10. J. Lemaitre, "A Course on Damage Mechanics," 2nd Edition, Springer, Berlin, 1996.

11. C. L. Chow and J. Wang, "An Anisotropic Theory of Continuum Damage Mechanics for Ductile Fracture," Engineering Fracture Mechanics, Vol. 27, No. 5, 1987, pp. 547-558.

12. W. Tai and B. Yang, "A New Damage Mechanics Criterion for Ductile Fracture," Engineering Fracture Mechanics, Vol. 27, No. 4, 1987, pp. 371-378.

13. T.-J. Wang, "Unified CDM Model and Local Criterion for Ductile Fracture: I—Unified CDM Model for Ductile Fracture," Engineering Fracture Mechanics, Vol. 42, 1992, pp. 177-183.

14. T.-J. Wang, "Unified CDM Model and Local Criterion for Ductile Fracture: II—Ductile Fracture Local Criterion Based on the CDM Model," Engineering Fracture Mechanics, Vol. 42, 1992, pp. 185-193.

15. S. Chandrakanth and P. Pandey, "An Isotropic Damage Model for Ductile Material," Engineering Fracture Mechanics, Vol. 50, No. 4, 1995, pp. 457-465.

16. N. Bonora, "A Nonlinear CDM Model for Ductile Failure," Engineering Fracture Mechanics, Vol. 58, No. 1, 1997, pp. 11-28.

17. L. Storojeva, D. Ponge, R. Kaspar and D. Raabe, "Development of Microstructure and Texture of Medium Carbon Steel during Heavy Warm Deformation," Acta Materialia, Vol. 52, No. 8, 2004, pp. 2209-2220.

18. F. A. McClintock, "A Criterion for Ductile Fracture by the Growth of Holes," Journal of Applied Mechanics, Vol. 35, No. 2, 1968, pp. 363-371.

19. F. Sidoroff, "On the Formulation of Plasticity and Viscoplasticity with Internal Variables," Archiwum Mechaniki Stosowanej, 1975, pp. 807-819.

20. Simulia Abaqus 6.10, "User Subroutines Reference Manual," 2010.

21. S. W. Lee and F. Pourboghrat, "Finite Element Simulation of the Punchless Piercing Process with Lemaitre Damage Model," International Journal of Mechanical Sciences, Vol. 47, No. 11, 2005, pp. 1756-1768.

22. P. Ludwik, "Elements der Technologischen Mechanik," 3rd Edition, Vol. 32, Springer, Berlin, 1909.

23. J. A. Benito, J. M. Manero, J. Jorba and A. Roca, "Change of Young's Modulus of Cold-Deformed Pure Iron in a Tensile Test," Metallurgical and Materials Transactions A, Vol. 36A, 2005, pp. 3317-3324.

24. D. J. Celentano and J.-L. Chaboche, "Experimental and Numerical Characterization of Damage Evolution in Steels," International Journal of Plasticity, Vol. 23, No. 10-11, 2007, pp. 1739-1762.

25. D. J. Celentano, P. E. Tapia and J. L. Chaboche, "Experimental and Numerical Characterization If Damage Evolution in Steels," In: G. Buscaglia, E. Dari and O. Zamonsky, Eds., Mecánica Computacional, Vol. XXIII, Bariloche, Argentina, 2004, pp. 45-58.

26. S. P. Tsiloufas, "Estudo da Fratura Dúctil em Chapas de Médio Carbono Sob a Ótica da Teoria da Mecânica do Dano (in Portuguese)," MSc. Dissertation, University of São Paulo, São Paulo, 2012.

27. J. Gurland, "Observations on the Fracture of Cementite Particles in a Spheroidized 1.05% C Steel Deformed at Room Temperature," Acta Metallurgica, Vol. 20, No. 5, 1972, pp. 735-741.

CHAPTER 3

Some Thermodynamic Problems in Continuum Mechanics

Zhen-Bang Kuang

Shanghai Jiaotong University, Shanghai, China

1. INTRODUCTION

Classical thermodynamics discusses the thermodynamic system, its surroundings and their common boundary. It is concerned with the state of thermodynamic systems at equilibrium, using macroscopic, empirical properties directly measurable in the laboratory (Wang, 1955; Yunus, Michael and Boles, 2011). Classical thermodynamics model exchanges of energy, work and heat based on the laws of thermodynamics. The first law of thermodynamics is a principle of conservation of energy and defines a specific internal energy which is a state function of the system. The second law of thermodynamics is a principle to explain the irreversibile phenomenon in nature. The entropy of an isolated non-equilibrium system will tend to increase over time, approaching a maximum value at equilibrium. Thermodynamic laws are generally valid and can be applied to systems about which only knows the balance of energy and matter transfer. The thermodynamic state of the system can be described by a number of state variables. In continuum mechanics state variables usually are pressurep, volumeV, stressσ, strainε, electric field strengthE, electric displacementD, magnetic induction densityB, magnetic field strengthH, temperatureT, entropy per volumes, chemical potential per volume μand concentration c respectively. Conjugated variable pairs are(p,V),(σ,ε),(E,D),(H,B),(T,S),(μ,c). There is a convenient and useful combination system in continuum mechanics: variables V,ε,E,H,T,μ are used as independent variables and variables p,σ,D,B,S,c are used as dependent variables. In this chapter we only use these conjugated variable pairs, and it is easy to extend to other conjugated variable pairs. In the later discussion we only use the following thermodynamic state functions: the internal energy U and the electro-magneto-chemical Gibbs free energy geμ(ε,E,H,T,μ) per volume in an electro-magneto-elastic material. They are taken as

$$\begin{aligned} &\mathrm{d}\mathcal{U}(\boldsymbol{\varepsilon},\boldsymbol{D},\boldsymbol{B},s,c)=\boldsymbol{\sigma}:\mathrm{d}\varepsilon+\boldsymbol{E}\cdot\mathrm{d}\boldsymbol{D}+\boldsymbol{H}\cdot\mathrm{d}\boldsymbol{B}+T\mathrm{d}s+\mu\mathrm{d}c\ ;\quad \boldsymbol{\sigma}:\mathrm{d}\varepsilon=\sigma_{ij}\mathrm{d}\varepsilon_{ij}\\ &\mathrm{d}g_{eu}(\boldsymbol{\varepsilon},\boldsymbol{E},\boldsymbol{H},T,\mu)=\mathrm{d}(\mathcal{U}-\boldsymbol{E}\cdot\boldsymbol{D}-\boldsymbol{H}\cdot\boldsymbol{B}-Ts-\mu c)=\boldsymbol{\sigma}:\mathrm{d}\varepsilon-\boldsymbol{D}\cdot\mathrm{d}\boldsymbol{E}-\boldsymbol{B}\cdot\mathrm{d}\boldsymbol{H}-s\mathrm{d}T-c\mathrm{d}\mu \end{aligned} \tag{1}$$

Other thermodynamic state functions and their applications can be seen in many literatures (Kuang, 2007, 2008a, 2008b, 2009a, 2009b, 2010, 2011a, 2011b). For the case without chemical potential geμ=ge is the electromagnetic Gibbs free energy. For the case without electromagnetic field geμ=gμ is the Gibbs free energy with chemical potential. For the case without chemical potential and electromagnetic field geμ=g is the Helmholtz free energy.

In this chapter two new problems in the continuum thermodynamics will be discussed. The first is that in traditional continuum thermodynamics including the non-equilibrium theory the dynamic effect of the temperature is not fully considered. When the temperature T is varied, the extra heat or entropy should be input from the environment. When c is varied, the extra chemical potential μ is also needed. So the general inertial entropy theory (Kuang, 2009b, 2010) is introduced into the continuum thermodynamics. The temperature and diffusion waves etc. with finite phase velocity can easily be obtained from this theory. The second is that usually we consider the first law only as a conservation law of different kinds of energies, but we found that it is also containing a physical variational principle, which gives a true process for all possible process satisfying the natural constrained conditions (Kuang, 2007, 2008a, 2008b, 2009a 2011a, 2011b). Introducing the physical variational principle the governing equations in continuum mechanics and the general Maxwell stress and other theories can naturally be obtained. When write down the energy expression, we get the physical variational principle immediately and do not need to seek the variational functional as that in the usual mathematical methods. The successes of applications of these theories in continuum mechanics are indirectly prove their rationality, but the experimental proof is needed in the further.

2. INERTIAL ENTROPY THEORY

2.1. Basic theory in linear thermoelastic material

In this section we discuss the linear thermoelastic material without chemical reaction, so in Eq. (1) the term $-D \cdot dE - B \cdot dH - c d\mu$ is omitted. It is also noted that in this section the general Maxwell stress is not considered. The classical thermodynamics discusses the equilibrium system, but when extend it to continuum mechanics we need discuss a dynamic system which is slightly deviated from the equilibrium state. In previous literatures one point is not attentive that the variation of temperature should be supplied extra heat from the environment. Similar to the inertial force in continuum mechanics we modify the thermodynamic entropy equation by adding a term containing an inertial heat or the inertial entropy (Kuang, 2009b), i.e.

$$T\dot{s} + T\dot{s}^{(a)} = \dot{r} - q_{i,i} = \dot{r} - (T\dot{\eta}_i)_{,i}, \quad \dot{s}^{(a)} = \rho_s \ddot{T}, \quad s^{(a)} = \rho_s \dot{T} = \rho_{s0}(C/T_0)\dot{T}$$
$$\dot{s} = \dot{s}^{(r)} + \dot{s}^{(i)}; \quad \dot{s}^{(r)} + \dot{s}^{(a)} = \dot{r}/T - \dot{\eta}_{i,i}; \quad \dot{\boldsymbol{\eta}} = \boldsymbol{q}/T$$
$$T\dot{s}^{(i)} = T\dot{s} - T\dot{s}^{(r)} = T\dot{s} + T\dot{s}^{(a)} - \dot{r} + T\dot{\eta}_{i,i} = -\dot{\eta}_i T_{,i} \geq 0; \quad \dot{s}^{(i)} = -\dot{\eta}_i T_{,i}/T \tag{2}$$

where s(a) is called the reversible inertial entropy corresponding to the inertial heat; ρsis called the inertial entropy coefficient, ρs0is also a constant having the dimension of the time; sis the entropy saved in the system, ˙s(r)and ˙s(i) are the reversible and irreversible parts of thes, T˙sis the absorbed heat rate of the system from the environment, T˙s(a)=ρsT¨Tis the inertial heat rate and ˙s(a) is proportional to the acceleration of the temperature; ris the external heat source strength, qis the heat flow vector per interface area supplied by the environment, ηis the entropy displacement vector, ˙ηis the entropy flow vector. Comparing Eq. (2) with the classical entropy equation it is found that in Eq. (2) we use T˙s+T˙s(a) to instead of T˙s in the classical theory. In Eq. (2) sis still a state function because s(a) is reversible. As in classical theory the dissipative energy ˙h′ and its Legendre transformation or “the complement dissipative energy” hare respectively

$$\dot{h}' = \mathrm{d}\dot{h}'/\mathrm{d}t = T\dot{s}^{(i)} = -T_{,i}\dot{\eta}_i, \quad \dot{h} = -T_{,i}\dot{\eta}_i + \left(T_{,i}\eta_i\right)^{\bullet} = \eta_i\dot{T}_{,i} \tag{3}$$

Using the theory of the usual irreversible thermodynamics (Groet, 1952; Gyarmati, 1970; Jou, Casas-Vzquez, Lebon, 2001; Kuang, 2002) from Eq. (3) we get

$$\begin{gathered} \dot{\eta}_i = \dot{\eta}_i(T_{,j}), \quad \text{or} \quad \dot{\eta}_i = -\lambda_{ij}T^{-1}T_{,j}, \quad T\dot{\eta}_i = q_i = -\lambda_{ij}T_{,j} \\ T_{,j} = -\hat{\lambda}_{ij}T\dot{\eta}_i = -\hat{\lambda}_{ij}q_i, \quad \hat{\lambda}_{ij} = \lambda_{ij}^{-1} \end{gathered} \tag{4}$$

where λ is the usual heat conductive coefficient. Eq. (4) is just the Fourier's law.

2.2. Temperature wave in linear thermoelastic material

The temperature wave from heat pulses at low temperature propagates with a finite velocity. So many generalized thermoelastic and thermopiezoelectric theories were proposed to allow a finite velocity for the propagation of a thermal wave. The main generalized theories are: Lord—Shulman theory (1967), Green—Lindsay theory (1972) and the inertial entropy theory (Kuang, 2009b).

In the Lord—Shulman theory the following Maxwell-Cattaneo heat conductive formula for an isotropic material was used to replace the Fourier's law, but the classical entropy equation is kept, i.e. they used

$$q_i + \tau_0 \dot{q}_i = -\lambda T_{,i}, \quad T\dot{s} = \dot{r} - q_{i,i} \tag{5}$$

where τ0 is a material parameter with the dimension of time. After linearization and neglecting many small terms they got the following temperature wave and motion equations for an isotropic material:

$$\begin{aligned} &\lambda T_{,ii} = C(\dot{T} + \tau_0 \ddot{T}) + \left[2G(1+\nu)/(1-2\nu)\right]\alpha T_0(\dot{\varepsilon}_{jj} + \tau_0 \ddot{\varepsilon}_{jj}) \\ &\left[G/(1-2\nu)\right]u_{j,ij} + Gu_{i.jj} - \left[2G(1+\nu)/(1-2\nu)\right]\alpha T_{,i} = \rho \ddot{u}_i \end{aligned} \tag{6}$$

where C is the specific heat, αis the thermal expansion coefficient, G and ν are the shear modulus and Poisson's ratio respectively. From Eq. (5) we can get

$$T\dot{s} - \tau_0 T\ddot{s} = \lambda T_{,ii} + \left(\dot{r} + \tau_0 \ddot{r}\right)$$

From above equation it is difficult to consider that s is a state function.

The Green—Lindsay theory with two relaxation times was based on modifying the Clausius-Duhemin inequality and the energy equation; In their theory they used a new temperature function ϕ(T,˙T) to replace the usual temperatureT. They used

$$\int_V \dot{s}\mathrm{d}V - \int_V (r/\phi)\mathrm{d}V + \int_a (q_i/\phi)n_i\mathrm{d}a \geq 0, \quad \phi = \phi(T,\dot{T}), \quad T = \phi(T,0)$$
$$g = \mathcal{U} - \phi s, \quad g = g\left(T,\dot{T},\varepsilon_{ij}\right) \tag{7}$$

After linearization and neglecting small terms, finally they get (here we take the form in small deformation for an isotropic material)

$$\lambda T_{,ii} = C(\dot{T} + \tau_0 \ddot{T}) + \gamma T_0 \dot{\varepsilon}_{jj}, \qquad \sigma_{ji,j} + \rho f_i = \rho \ddot{u}_i$$
$$\sigma_{ij} = \left[2G\nu/(1-2\nu)\right]\varepsilon_{kk}\delta_{ij} + 2G\varepsilon_{ij} - \gamma(\theta + \tau_1\dot{\theta}) \tag{8}$$

whereτ0, τ1and γ are material constants.

Now we discuss the inertial entropy theory (Kuang, 2009b). The Helmholtz free energy g and the complement dissipative energy h assumed in the form

$$g\,(\varepsilon_{kl},\vartheta) = (1/2)C_{ijkl}\varepsilon_{ji}\varepsilon_{lk} - \alpha_{ij}\varepsilon_{ij}\vartheta - (1/2T_0)C\vartheta^2$$
$$\delta h = -\left[\int_0^t \left(\lambda_{ij}/T\right)\vartheta_{,i}d\tau)\right]\delta\vartheta_{,j}, \quad \vartheta = T - T_0$$
$$C_{ijkl} = C_{jikl} = C_{ijlk} = C_{klij}, \quad \alpha_{ij} = \alpha_{ji}, \quad \lambda_{ij} = \lambda_{ji} \tag{9}$$

where T0 is the reference (or the environment) temperature, C_{ijkl},$\alpha_{ij}$are material constants. In Eq. (9a) it is assumed that s=0 when T=T0 orϑ=0. It is obvious thatT,j=ϑ,j, ˙T=˙ϑ.

The constitutive (or state) and evolution equations are

$$\sigma_{ij} = \partial g/\partial \varepsilon_{ij} = C_{ijkl}\varepsilon_{kl} - \alpha_{ij}\vartheta, \quad s = -\partial g/\partial \vartheta = \alpha_{ij}\varepsilon_{ij} + C\vartheta / T_0$$
$$\eta_i = \partial h/\partial \vartheta_{,i} = -\int_0^t \left(\lambda_{ij}/T\right)\vartheta_{,j}\mathrm{d}\tau, \quad T\dot{\eta}_i = q_i = -\lambda_{ij}\vartheta_{,j} \tag{10}$$

Using Eq. (10), Eq. (9a) can be rewritten as

$$g = (1/2)C_{ijkl}\varepsilon_{ji}\varepsilon_{lk} + g^{(T)}, \quad g^{(T)} = -(1/2)s\vartheta - (1/2)\alpha_{ij}\varepsilon_{ij}\vartheta \tag{9b}$$

where $g^{(T)}$ is the energy containing the effect of the to temperature.

Substituting the entropy s and T˙ηi in Eq. (10) and s(a) in (2) into $T\dot{s} + T\dot{s}^{(a)} = \dot{r} - (T\dot{\eta}_i)_{,i}$ in Eq. (2) we get

$$T\left(\alpha_{ij}\varepsilon_{ij} + C\vartheta / T_0\right)^{\bullet} + \rho_s T\ddot{\vartheta} = \dot{r} + \left(\lambda_{ij}\vartheta_{,j}\right)_{,i} \tag{11}$$

When material coefficients are all constants from (11) we get

$$\rho_s T\ddot{\vartheta} + CT\dot{\vartheta} / T_0 - \lambda_{ij}\vartheta_{,ji} = \dot{r} - \alpha_{ij}T\dot{\varepsilon}_{ij} \tag{12a}$$

Eq. (12a) is a temperature wave equation with finite phase velocity. For an isotropic elastic material and the variation of the temperature is not large, from Eq. (12a) we get

$$(C/T_0)(\rho_{s0}\ddot{\vartheta}+\dot{\vartheta})-(\lambda/T)\vartheta_{,ii}=\dot{r}/T-\alpha\dot{\varepsilon}_{ii} \quad or$$
$$\lambda\vartheta_{,ii}=C(\rho_{s0}\ddot{\vartheta}+\dot{\vartheta})+\alpha T_0\dot{\varepsilon}_{ii}-\dot{r} \tag{12b}$$

Comparing the temperature wave equation Eq. (12b) with the Lord—Shulman theory (Eq. (6)) it is found that in Eq. (12b) a term τ0¨εjj is lacked (in different notations),but with that in the Green—Lindsay theory (Eq. (8)) is similar (in different notations). For the purely thermal conductive problem three theories are fully the same in mathematical form.

The momentum equation is

$$\sigma_{ij,j}+f_i=\rho\ddot{u}_i \tag{13}$$

where f is the body force per volume, ρis the density. Substituting the stress σ in Eq. (10) into (13) we get

$$(C_{ijkl}\varepsilon_{kl}-\alpha_{ij}\vartheta)_{,j}+f_i=\rho\ddot{u}_i, \quad or \quad \rho\ddot{u}_i=C_{ijkl}u_{k,lj}-\alpha_{ij}\vartheta_{,j}+f_i \tag{14}$$

Comparing the elastic wave equation Eq. (14) with the Green—Lindsay theory (Eq. (8)) it is found that in Eq. (14) a term γτ1˙ϑ,i is lacked (in different notations), but with the Lord—Shulman theory (Eq. (6)) is similar (in different notations).

2.3. Temperature wave in linear thermo - viscoelastic material

In the pyroelectric problem (without viscous effect) through numerical calculations Yuan and Kuang (2008, 2010) pointed out that the term containing the inertial entropy attenuates the temperature wave, but enhances the elastic wave. For a given material there is a definite value ofρ∗s0, when ρs0>ρ∗s0 the amplitude of the elastic wave will be increased with time. For BaTio3ρ∗s0 is about10−13s. In the Lord—Shulman theory critical value τ∗0is about10−8s. In order to substantially eliminate the increasing effect of the amplitude of the elastic wave the viscoelastic effect is considered as shown in this section.

Using the irreversible thermodynamics (Groet, 1952; Kuang, 1999, 2002) we can assume

$$
\begin{aligned}
&g=(1/2)C_{ijkl}\varepsilon_{ji}\varepsilon_{lk}-\alpha_{ij}\varepsilon_{ij}\vartheta-(1/2T_0)C\vartheta^2\\
&\sigma_{ij}^{(r)}=\partial g/\partial\varepsilon_{ij}=C_{ijkl}\varepsilon_{kl}-\alpha_{ij}\vartheta,\quad s=-\partial g/\partial\vartheta=\alpha_{ij}\varepsilon_{ij}+C\vartheta/T_0\\
&\delta h=\beta_{ijkl}\dot{\varepsilon}_{ji}\delta\dot{\varepsilon}_{lk}+\eta_j\delta\vartheta_{,j}=\beta_{ijkl}\dot{\varepsilon}_{ji}\delta\dot{\varepsilon}_{lk}-\left[\int_0^t(\lambda_{ij}/T)\vartheta_{,i}d\tau)\right]\delta\vartheta_{,j},\\
&\sigma_{ij}^{(i)}=\partial h/\partial\dot{\varepsilon}_{ij}=\beta_{ijkl}\dot{\varepsilon}_{kl},\ \eta_i=\partial h/\partial\vartheta_{,i}=-\int_0^t(\lambda_{ij}/T)\vartheta_{,j}\mathrm{d}\tau,\ T\dot{\eta}_i=q_i=-\lambda_{ij}\vartheta_{,j}\\
&\sigma_{ij}=\sigma_{ij}^{(r)}+\sigma_{ij}^{(i)}=C_{ijkl}\varepsilon_{kl}+\beta_{ijkl}\dot{\varepsilon}_{kl}-\alpha_{ij}\vartheta
\end{aligned}
\tag{15}
$$

where $\sigma_{ij}^{(r)}$ and $\sigma_{ij}^{(r)}$ are the reversible and irreversible parts of the stressσij,˙εij=dεij/dt. Comparing Eqs. (9) and (10) with (15) it is found that only a term $\beta_{ijkl}\dot{\varepsilon}_{ji}\delta\dot{\varepsilon}_{lk}$ is added to the rate of the complement dissipative energy in Eq. (15). Substituting the entropy s and T˙ηi in Eq. (15) and s(a) in (2) into T˙s+T˙s(a)=˙r−(T˙ηi),i in Eq. (2) we still get the same equation (12).

Substituting the stress σ in Eq. (15) into (13) we get

$$\left(C_{ijkl}\varepsilon_{kl} + \beta_{ijkl}\dot{\varepsilon}_{kl} - \alpha_{ij}\vartheta\right)_{,j} + f_i = \rho\ddot{u}_i, \quad or \quad \rho\ddot{u}_i = C_{ijkl}u_{k,lj} + \beta_{ijkl}\dot{u}_{k,lj} - \alpha_{ij}\vartheta_{,j} + f_i \tag{16}$$

In one dimensional problem for the isotropic material from Eq. (15) we have

$$\sigma = Y\varepsilon + \beta\dot{\varepsilon} - \alpha\vartheta, \quad s = \alpha\varepsilon + C\vartheta / T_0 \tag{17}$$

where Y is the elastic modulus, βis a viscose coefficient, αis the temperature coefficient.

When there is no body force and body heat source, Eqs. (12) and (16) are reduced to

$$\begin{aligned} &C\left(\rho_{s0}\ddot{\vartheta} + \dot{\vartheta}\right) - \lambda\vartheta'' + \alpha T_0\dot{u}' = 0 \\ &\rho\ddot{u} - Yu'' - \beta\dot{u}'' + \alpha\vartheta' = 0 \end{aligned} \tag{18}$$

where $\dot{\varphi}=\partial\varphi/\partial t, \varphi'=\partial\varphi/\partial x$ for any function φ. For a plane wave propagating along direction *x* we assume

$$u = U\exp\left[\mathrm{i}(kx-\omega t)\right], \quad \theta = \Theta\exp\left[\mathrm{i}(kx-\omega t)\right] \tag{19}$$

where U,Θ are the amplitudes of u and ϑ respectively, kis the wave number and ω is the circular frequency. Substituting Eq. (19) into (18) we obtain

$$\begin{aligned} &\left[(Y-\mathrm{i}\beta\omega)k^2-\rho\omega^2\right]U+\mathrm{i}\alpha k\Theta=0 \\ &\alpha T_0 k\omega U+\left[\lambda k^2-C\left(\rho_{s0}\omega^2+\mathrm{i}\omega\right)\right]\Theta=0 \end{aligned} \tag{20}$$

In order to have nontrivial solutions forU,Θ, the coefficient determinant of Eq. (20) should be vanished:

$$\begin{vmatrix} (Y-\mathrm{i}\beta\omega)k^2-\rho\omega^2 & \mathrm{i}\alpha k \\ \alpha T_0 k\omega & \lambda k^2-C\left(\rho_{s0}\omega^2+\mathrm{i}\omega\right) \end{vmatrix} = \begin{vmatrix} ak^2-\rho\omega^2 & \mathrm{i}\alpha k \\ \alpha T_0 k\omega & \lambda k^2-Cb \end{vmatrix} = 0 \tag{21a}$$

Where

$$\begin{aligned} &a = Y-\mathrm{i}\beta\omega = r_Y e^{\mathrm{i}\theta_Y}, \quad r_Y=\sqrt{a\bar{a}}=\sqrt{Y^2+\beta^2\omega^2}, \quad \sin\theta_Y=-\beta\omega/r_Y \\ &b=\rho_{s0}\omega^2+\mathrm{i}\omega=r_T e^{\mathrm{i}\theta_T}, \quad r_T=\sqrt{b\bar{b}}=\omega\sqrt{\rho_{s0}^2\omega^2+1}, \quad \sin\theta_T=\omega/r_T \end{aligned} \tag{21b}$$

From Eq. (21) we get

$$\lambda ak^4 - \left(Cab + \lambda\rho\omega^2 + i\alpha^2 T_0\omega\right)k^2 + \rho\omega^2 Cb = 0$$

$$k = \left\{ \frac{1}{2\lambda r_Y}\left[\left(Cr_Y r_T e^{i\theta_T} + \lambda\rho\omega^2 e^{-i\theta_Y} + i\alpha^2 T_0\omega e^{-i\theta_Y}\right) \pm \sqrt{\left(Cr_Y r_T e^{i\theta_T} - \lambda\rho\omega^2 e^{-i\theta_Y} + i\alpha^2 T_0\omega e^{-i\theta_Y}\right)^2 + 4i\alpha^2 T_0\omega r_Y r_T e^{i(\theta_T-\theta_Y)}}\right]\right\}^{1/21/2}$$

(22)

where the symbol "+" is applied to the wave number kT of the temperature wave and the symbol "−" is applied to the wave number of the viscoelastic wavekY. If the temperature wave does not couple with the elastic wave, then αis equal to zero. In this case we have

$$k = \sqrt{\left(2\lambda r_Y\right)^{-1}\left\{\left(Cr_T r_Y e^{i\theta_T} + \lambda\rho\omega^2 e^{-i\theta_Y}\right) \pm \left(Cr_T r_Y e^{i\theta_T} - \lambda\rho\omega^2 e^{-i\theta_Y}\right)\right\}}$$

$$k_Y = \omega\sqrt{\rho/r_Y}e^{-i\theta_Y/2}, \quad k_T = \sqrt{Cr_T/\lambda}e^{i\theta_T/2}$$

(23)

Because −θY>0due to β>0 and θT>0 due tops0>0, a pure viscoelastic wave or a pure temperature waves is attenuated. The pure elastic wave does not attenuate due toβ=0.

For the general case in Eq. (22) a coupling term iα2T0ωk2 is appeared. It is known that

$$\left|\mathrm{Im}\left(Cr_Y r_T e^{i\theta_T} + \lambda\rho\omega^2 e^{-i\theta_Y} + i\alpha^2 T_0\omega e^{-i\theta_Y}\right)\right| > \left|\mathrm{Im}\left(Cr_Y r_T e^{i\theta_T} - \lambda\rho\omega^2 e^{-i\theta_Y} + i\alpha^2 T_0\omega e^{-i\theta_Y}\right)\right|$$

It means that ImkT>0 or the temperature wave is always an attenuated wave. If

$$\operatorname{Im}\left[\left(Cr_Y r_T e^{i\theta_T} - \lambda\rho\omega^2 e^{-i\theta_Y} + i\alpha^2 T_0 \omega e^{-i\theta_Y}\right)^2 + 4i\alpha^2 T_0 \omega r_Y r_T e^{i(\theta_T - \theta_Y)}\right]$$
$$< \operatorname{Im}\left(Cr_Y r_T e^{i\theta_T} + \lambda\rho\omega^2 e^{-i\theta_Y} + i\alpha^2 T_0 \omega e^{-i\theta_Y}\right)^2$$

(24)

we get ImkY>0 or in this case the elastic wave is an attenuated wave, otherwise is enhanced.

Introducing the viscoelastic effect in the elastic wave as shown in this section can substantially eliminate the increasing effect of the amplitude of the elastic wave with time.

2.4. Temperature wave in thermo-electromagneto-elastic material

In this section we discuss the linear thermo-electromagneto-elastic material without chemical reaction and viscous effect, so the electromagnetic Gibbs free energy ge in Eq. (1) should keep the temperature variable. The electromagnetic Gibbs free energy ge and the complement dissipative energy he in this case are assumed respectively in the following form

$$g_e(\varepsilon_{kl}, E_k, H_k, \vartheta) = (1/2)C_{ijkl}\varepsilon_{ji}\varepsilon_{lk} - e^e_{kij}E_k\varepsilon_{ij} - (1/2)\epsilon_{ij}E_iE_j - \tau^e_i E_i\vartheta$$
$$-e^m_{kij}H_k\varepsilon_{ij} - (1/2)\mu_{ij}H_iH_j - \tau^m_i H_i\vartheta - \alpha_{ij}\varepsilon_{ij}\vartheta - (1/2T_0)C\vartheta^2$$
$$\delta h_e = -(\int_0^t (\lambda_{ij}/T)\vartheta_{,i}d\tau)\delta\vartheta_{,j}\ (= \eta_j\delta\vartheta_{,j}), \quad \vartheta = T - T_0$$
$$C_{ijkl} = C_{jikl} = C_{ijlk} = C_{klij},\ e^e_{kij} = e^e_{kji},\ \epsilon_{kl} = \epsilon_{lk},\ e^m_{kij} = e^m_{kji},\ \mu_{kl} = \mu_{lk},\ \alpha_{ij} = \alpha_{ji}$$

(25)

where $e^e_{kij}, \epsilon_{kl}, \tau^e_i, e^m_{kij}, \mu_{kl}, \tau^m_i$ are material constants. The constitutive equations are

$$\sigma_{ij} = C_{ijkl}\varepsilon_{kl} - e^e_{kij}E_k - e^m_{kij}H_k - \alpha_{ij}\vartheta,\ D_i = \epsilon_{ij}E_j + e^e_{ijk}\varepsilon_{jk} + \tau^e_i\vartheta$$
$$B_i = \mu_{ij}H_j + e^m_{ijk}\varepsilon_{jk} + \tau^m_i\vartheta,\ s = \alpha_{ij}\varepsilon_{ij} + \tau^e_i E_i + \tau^m_i H_i + C\vartheta / T_0$$
$$\eta_i = \partial h_v/\partial\vartheta_{,i} = -\int_0^t (\lambda_{ij}/T)\vartheta_{,j}\mathrm{d}\tau,\ T\dot{\eta}_i = q_i = -\lambda_{ij}\vartheta_{,j} \quad (26)$$

Similar to derivations in sections 2.2 and 2.3 it is easy to get the governing equations:

$$T\left(\alpha_{ij}\varepsilon_{ij} + \tau^e_i E_i + \tau^m_i H_i + C\vartheta / T_0\right)^{\boldsymbol{\cdot}} + \rho_s T\ddot{T} = \dot{r} + \left(\lambda_{ij}\vartheta_{,j}\right)_{,i} \quad (27)$$

$$\left(C_{ijkl}\varepsilon_{kl} - e^e_{kij}E_k - e^m_{kij}H_k - \alpha_{ij}\vartheta\right)_{,j} + f_i = \rho\ddot{u}_i,$$
$$\left(\epsilon_{ij}E_j + e^e_{kij}\varepsilon_{kl} + \tau^e_i\vartheta\right)_{,i} = \rho_e,\quad \left(\mu_{ij}H_j + e^m_{kij}\varepsilon_{kl} + \tau^m_i\vartheta\right)_{,i} = 0 \quad (28)$$

where ρe is the density of the electric charge. The boundary conditions are omitted here.

2.5. Thermal diffusion wave in linear thermoelastic material

The Gibbs equation of the classical thermodynamics with the thermal diffusion is:

$$
\begin{aligned}
&T\dot{s} = \dot{r} - q_{i,i} + \mu d_{i,i}, \quad T\dot{s} + \mu\dot{c} = \dot{r} - q_{i,i} = \dot{r} - \left(T\dot{\eta}_i\right)_{,i}, \quad d_{i,i} = -\dot{c} \\
&\dot{\mathcal{U}} = \boldsymbol{\sigma} : \dot{\boldsymbol{\varepsilon}} + T\dot{s} + \mu\dot{c}, \quad \dot{g}_\mu = \boldsymbol{\sigma} : \dot{\boldsymbol{\varepsilon}} - s\dot{T} - c\dot{\mu}
\end{aligned}
\tag{29}
$$

where μis the chemical potential, dis the flow vector of the diffusing mass, cis the concentration. In discussion of the thermal diffusion problem we can also use the free energy $\dot{g}_c = \boldsymbol{\sigma} : \dot{\boldsymbol{\varepsilon}} - s\dot{T} + \mu\dot{c}$ (Kuang, 2010), but here it is omitted. Using relations

$$
T^{-1}q_{i,i} = \left(T^{-1}q_i\right)_{,i} + T^{-2}q_iT_{,i}, \quad T^{-1}\mu d_{i,i} = \left(T^{-1}\mu d_i\right)_{,i} - d_i\left(T^{-1}\mu\right)_{,i}
$$

From Eq. (29) (Kuang, 2010) we get:

$$
\begin{aligned}
&\dot{s} = \dot{s}^{(r)} + \dot{s}^{(i)}; \quad T\dot{s}^{(r)} = \dot{r} - T\left(q_i/T - \mu d_i/T\right)_{,i} \\
&T\dot{s}^{(i)} = T\dot{s} - T\dot{s}^{(r)} = -T_{,i}\dot{\eta}_i - \mu'_{,i}\dot{\xi}'_i = \geq 0, \quad \mu'_{,i} = \left(\mu/T\right)_{,i}, \quad \dot{\xi}'_i = Td_i
\end{aligned}
\tag{30}
$$

where T˙s(i) is the irreversible heat rate. According to the linear irreversible thermodynamics the irreversible forces are proportional to the irreversible flow (Kuang, 2010; Gyarmati, 1970; De Groet, 1952), we can write the evolution equations in the following form

$$T\dot{\eta}_i = -\lambda_{ij}(T)T_{,i} - L_{ij}(T)T\mu'_{,i}, \quad T^{-1}\dot{\xi}'_i = -D_{ij}(T)T\mu'_{,i} - L_{ij}(T)T_{,i} \tag{31a}$$

where Dij is the diffusing coefficients and Lij is the coupling coefficients. The linear irreversible thermodynamics can only give the general form of the evolution equation, the concrete exact formula should be given by experimental results. Considering the experimental facts and the simplicity of the requirement for the variational formula, when the variation of T is not too large, Eq. (31a) can also be approximated by

$$\begin{aligned}
&T\dot{s}^{(i)} \approx -T_{,i}\dot{\eta}_i - \mu_{,i}\dot{\xi}_i \geq 0; \quad \dot{\xi}_i = d_i \\
&T\dot{\eta}_i = -\lambda_{ij}(T)T_{,i} - L_{ij}(T)\mu_{,i}, \quad \dot{\xi}_i = -D_{ij}(T)\mu_{,i} - L_{ij}(T)T_{,i} \\
&T_{,i} = -\hat{\lambda}_{ij}(T)T\dot{\eta}_i - \hat{L}_{ij}(T)\dot{\xi}_i, \quad \mu_{,i} = -\hat{D}_{ij}(T)\dot{\xi}_i - \hat{L}_{ij}(T)T\dot{\eta}_i
\end{aligned} \tag{31b}$$

Especially the coefficients $\lambda_{ij}, L_{ij}, D_{ij}, \hat{\lambda}_{ij}, \hat{L}_{ij}, \hat{D}_{ij}$ in Eq. (31b) can all be considered as symmetric constants which are adopted in following sections. Eq. (31) is the extension of the Fourier's law and Fick's law.

Eq. (29) shows that in the equation of the heat flow the role of T˙sis somewhat equivalent toμ˙c. So analogous to the inertial entropy s(a) we can also introduce the inertial concentration c(a) and introduce a general inertial entropy theory of the thermal diffusion problem. Eq.

(29) in the general inertial entropy theory is changed to (Kuang, 2010)

$$T\left(\dot{s}+\dot{s}^{(a)}\right)+\mu\left(\dot{c}+\dot{c}^{(a)}\right)=\dot{r}-q_{i,i}=\dot{r}-\left(T\dot{\eta}_i\right)_{,i};\quad \dot{c}+\dot{c}^{(a)}=-d_{i,i}$$
$$s^{(a)}=\int_0^t \dot{s}^{(a)}\mathrm{d}\tau,\quad \dot{s}^{(a)}=\rho_s\ddot{T};\quad c^{(a)}=\int_0^t \dot{c}^{(a)}\mathrm{d}\tau,\quad \dot{c}^{(a)}=\rho_c\ddot{\mu} \tag{32}$$

where ρc is the inertial concentration coefficient. Applying the irreversible thermodynamics we can get the Gibbs free energy gμ and the complement dissipative energy hμ as

$$g_\mu\left(\varepsilon_{kl},\vartheta,\mu\right)=\frac{1}{2}C_{ijkl}\varepsilon_{ji}\varepsilon_{lk}-\alpha_{ij}\varepsilon_{ij}\vartheta-\frac{1}{2T_0}C\vartheta^2-\frac{1}{2}b\mu^2-b_{ij}\varepsilon_{ij}\mu-a\mu\vartheta$$
$$\delta h_\mu=-T_{,i}\delta\eta_i-\mu_{,i}\delta\xi_i+\delta\left(T_{,i}\eta_i\right)+\delta\left(\mu_{,i}\xi_i\right)=\eta_j\delta\vartheta_{,j}+\xi_j\delta\mu_{,j}$$
$$=-\delta\vartheta_{,j}\int_0^t\left(\lambda_{ij}T^{-1}\vartheta_{,i}+L_{ij}T^{-1}\mu_{,i}\right)d\tau-\delta\mu_{,j}\int_0^t\left(L_{ij}\vartheta_{,i}+D_{ij}\mu_{,i}\right)d\tau \tag{33a}$$

where a,b,bij are also material constants. The constitutive and evolution equations are:

$$\sigma_{ij}=\partial g_\mu/\partial\varepsilon_{ij}=C_{ijkl}\varepsilon_{kl}-\alpha_{ij}\vartheta-b_{ij}\mu,$$
$$s=-\partial g_\mu/\partial\vartheta=\alpha_{ij}\varepsilon_{ij}+C\vartheta/T_0+a\mu,\quad c=-\partial g_\mu/\partial\mu=b\mu+b_{ij}\varepsilon_{ij}+a\vartheta$$
$$\eta_i=\partial h_\mu/\partial\vartheta_{,i}=-\int_0^t\left(\lambda_{ij}T^{-1}\vartheta_{,j}+L_{ij}T^{-1}\mu_{,j}\right)d\tau,\ \xi_i=\partial h_\mu/\partial\mu_{,i}=-\int_0^t\left(L_{ij}\vartheta_{,j}+D_{ij}\mu_{,j}\right)d\tau$$

(34)

Using Eq. (34) gμin Eq. (33a) can also be rewritten as

$$g_{\mu}(\varepsilon_{kl},\vartheta,\mu)=(1/2)C_{ijkl}\varepsilon_{ji}\varepsilon_{lk}+g_{\mu}^{(T)},\quad g_{\mu}^{(T)}=-(1/2)\left(s\vartheta+c\mu+\alpha_{ij}\varepsilon_{ij}\vartheta+b_{ij}\varepsilon_{ij}\mu\right)$$

(33b)

where $g_{\mu}^{(T)}$ is the energy containing the effects of temperature and concentration. Substituting Eq. (34) into Eq. (32) we get

$$T\left(\alpha_{ij}\varepsilon_{ij}+C\vartheta/T_0+a\mu\right)^{\bullet}+T\rho_s\ddot{\vartheta}$$
$$+\mu\left(b\mu+b_{ij}\varepsilon_{ij}+a\vartheta\right)^{\bullet}+\mu\rho_s\ddot{\vartheta}=\dot{r}+\left(\lambda_{ij}\vartheta_{,j}+L_{ij}\mu_{,j}\right)_{,i}$$
$$\left(b\mu+b_{ij}\varepsilon_{ij}+a\vartheta\right)^{\bullet}+\rho_c\ddot{\mu}=L_{ji}\vartheta_{,ij}+D_{ji}\mu_{,ij};\qquad \textit{In medium}\qquad (35)$$

If we neglect the term in second order μdi,i in Eq. (29), i.e. we take T˙s=˙r−qi,i and assume that T,I andμ,j are not dependent each other, i.e. in Eq. (31b) we assume $\dot{\eta}_i=-\lambda_{ij}T^{-1}T_{,j},\dot{\xi}_i=-D_{ij}\mu_{,j}$, then for˙r=0, Eq. (35) becomes

$$T\left(\alpha_{ij}\dot{u}_{i,j}+C\dot{\vartheta}/T_0+a\dot{\mu}+\rho_s\ddot{\vartheta}\right)=\lambda_{ij}\vartheta_{,j}$$
$$b\dot{\mu}+b_{ij}\dot{u}_{i,j}+a\dot{\vartheta}+\rho_c\ddot{\mu}=D_{ij}\mu_{,ji};\qquad \textit{In medium}\qquad (36)$$

The formulas in literatures analogous to Eq. (34) can be found, such as in Sherief, Hamza, and Saleh's paper (2004), where they used the Maxwell-Cattaneo formula.The momentum equation is

$$\left(C_{ijkl}u_{k,l}-\alpha_{ij}\vartheta-b_{ij}\mu\right)_{,j}+f_i=\rho\ddot{u}_i \tag{37}$$

The above theory is easy extended to more complex materials.

3. PHYSICAL VARIATIONAL PRINCIPLE

3.1. General theory

Usually it is considered that the first law of thermodynamics is only a principle of the energy conservation. But we found that the first law of thermodynamics is also a physical variational principle (Kuang, 2007, 2008a, 2008b, 2009a 2011a, 2011b). Therefore the first law of the classical thermodynamics includes two aspects: energy conservation law and physical variational principle:

$$\text{Classical Energy conservation:} \quad \int_V \mathrm{d}\mathcal{U}\mathrm{d}V-\mathrm{d}W-\mathrm{d}Q=0$$
$$\text{Classical physical variational principle:} \quad \delta\Pi=\int_V \delta\mathcal{U}\mathrm{d}V-\delta W-\delta Q=0 \tag{38}$$

where U is the internal energy per volume, *W* is the work applied on the body by the environment, Q is the heat supplied by the environment. According to Gibbs theory when the process is only slightly deviated from the equilibrium state dQ can be substituted by$\int$VTdsdV In practice we prefer to use the free energy

$$g = \mathcal{U} - Ts, \quad \mathrm{d}g = \mathrm{d}\mathcal{U} - s\mathrm{d}T - T\mathrm{d}s$$

Energy Principle: $\int_V \mathrm{d}g\mathrm{d}V - dW^* - \int_V s\mathrm{d}T\mathrm{d}V = 0$

Physical Variational Principle: $\delta\Pi = \int_V \delta g\mathrm{d}V - \delta W^* - \int_V s\delta T\mathrm{d}V = 0$ (39)

Here the physical variational principle is considered to be one of the fundamental physical law, which can be used to derive governing equations in continuum mechanics and other fields. We can also give it a simple explanation that the true displacement is one kind of the virtual displacement and obviously it satisfies the variational principle. Other virtual displacements cannot satisfy this variational principle, otherwise the first law is not objective. The physical variational principle is different to the usual mathematical variational method which is based on the known physical facts. In many problems the variation of a variable ϕ different with displacement , should be divided into local variation and migratory variation, i.e. the variationδϕ=δϕϕ+δuϕ , where the local variation δϕϕ of ϕ is the variation duo to the change of ϕ itself and the migratory variation δuϕ of ϕ is the variation of change of ϕ due to virtual displacements. In Eqs. (38) and (39) the new force produced by the migratory variation δuϕ will enter the virtual work δW or δW∗ as the same as the external mechanical force. But in the following sections we shall modify Eq. (39) or (38) to deal with this problem. The physical variational principle is inseparable with energy conservation law, so when the expressions of energies are given we get physical variational principle immediately. We need not to seek the variational functional as that in usual mathematical methods. In the following sections we show how to derive the governing equations with the general Maxwell stress of some kind of materials by using the physical variational principle. From this physical variational principle all of the governing equations in the continuum mechanics and physics can be carried out and this fact can be considered as the indirect evidence of the physical variational principle.

3.2. Physical variational principle in thermo-elasticity

In the thermo-elasticity it is usually considered that only the thermal process is irreversible, but the elastic process is reversible. So the free energy g and the complement dissipative energy can be assumed as that in Eq. (9). The corresponding constitutive and evolution equations are expressed in Eq. (10). As shown in section 3.1, the variation of the virtual temperature ϑ is divided into local variation $\delta\vartheta\vartheta$ due to the variation of ϑ itself and the migratory variation $\delta u\vartheta$ due to δu

$$\delta\vartheta = \delta_\vartheta\vartheta + \delta_u\vartheta, \quad \delta_u\vartheta = \vartheta_{,i}\delta u_i \tag{40}$$

In previous paper (Kuang, 2011a) we showed that the migratory variation of virtual electric and magnetic potentials will produce the Maxwell stress in electromagnetic media, which is also shown in section 3.4 of this paper. Similarly the migratory variation $\delta u\vartheta$will also produce the general Maxwell stress which is an external temperature stress. The effective general Maxwell stress can be obtained by the energy principle as that in electromagnetic media. Under assumptions that the virtual mechanical displacement u and the virtual temperature ϑ(orT) satisfy their own boundary conditions ui=u∗i,ϑ=ϑ∗ on auand aT respectively. The physical variational principle using the free energy in the inertial entropy theory for the thermo-elasticity can be expressed as:

$$\delta\Pi_T = \int_V \delta(g+h)\mathrm{d}V + \int_V g^{(T)}\delta u_{k,k}\mathrm{d}V - \delta Q^* - \delta W = 0$$

$$\delta Q^* = -\int_V\int_0^t (\dot{r}/T)\delta\vartheta\mathrm{d}\tau\mathrm{d}V - \int_V\int_0^t \dot{s}^{(i)}\delta\vartheta\mathrm{d}\tau\mathrm{d}V + \int_{a_q}\int_0^t \dot{\eta}^*\delta\vartheta\mathrm{d}\tau\mathrm{d}a + \int_V\int_0^t \rho_s\ddot{\vartheta}\delta\vartheta\mathrm{d}\tau\mathrm{d}V$$

$$\delta W = \int_V (f_k - \rho\ddot{u}_k)\delta u_k dV + \int_{a_\sigma} T_k^*\delta u_k da$$

(41)

where fk,T∗k and ˙η∗=˙η∗ini are the given mechanical body force, surface traction and surface entropy flow respectively. Eq. (41) is an alternative form of Eq. (39). In Eq. (41) the term $-\int_0^t \dot{s}^{(i)}\hat{\delta}\vartheta\mathrm{d}\tau = \int_0^t \eta_i\hat{\delta T}_{,i}\mathrm{d}\tau$ is the complement dissipative heat rate per volume corresponding to the inner complement dissipation energy rateδh. The entropy sincludes the contribution of– $-\int_0^t \dot{s}^{(\hat{i})}\delta\vartheta\mathrm{d}\tau$. The fact that the complement dissipation energy rate ∫VδhdV in δΠT and the internal irreversible complement heat rate $-\int_V\int_0^t \dot{s}^{(i)}$ in δQ∗ are simultaneously included in Eq. (41) allows us to get the temperature wave equation and the boundary condition of the heat flow from the variational principle. In Eq. (41) there are two kinds of variational formulas. The first is∫VδgdV+∫Vg(T)δuk,kdV−δW , in which the integrands contain variables themselves. The second is∫VδhdV−δQ∗, in which the integrands contain the time derivatives of variables, so it needs integrate with time t. This is the common feature of the irreversible process because in the irreversible process the integral is dependent to the integral path. It is noted that

$$\begin{aligned}
\int_V \delta g dV &= \int_V (C_{ijkl}\varepsilon_{kl} - \alpha_{ij}\vartheta)\delta u_{i,j} dV - \int_V (\alpha_{ij}\varepsilon_{ij} + C\vartheta / T_0)\delta\vartheta dV \\
&= \int_a \sigma_{ij} n_j \delta u_i da - \int_V \sigma_{ij,j} \delta u_i dV - \int_V s\delta\vartheta dV \\
\int_V g^{(T)} \delta u_{k,k} \mathrm{d}V &= -(1/2)\int_V \left(s + \alpha_{ij}\varepsilon_{ij}\right)\vartheta \delta u_{k,k} \mathrm{d}V \\
&= -(1/2)\int_a \left(s + \alpha_{ij}\varepsilon_{ij}\right)\vartheta n_k \delta u_k \mathrm{d}V + (1/2)\int_V \left[\left(s + \alpha_{ij}\varepsilon_{ij}\right)\vartheta\right]_{,k} \delta u_k \mathrm{d}V \\
\int_V \delta h dV &= -\int_a \left(\int_0^t \lambda_{ij} T^{-1} \vartheta_{,i} n_j \mathrm{d}\tau\right)\delta\vartheta da + \int_V \left[\int_0^t (\lambda_{ij} T^{-1} \vartheta_{,i})_{,j} \mathrm{d}\tau\right]\delta\vartheta dV
\end{aligned} \tag{42}$$

Finishing the variational calculation, we have

$$\begin{aligned}
\delta\Pi &= \int_{a_\sigma} (\tilde{\sigma}_{ij} n_j - T_i^*)\delta u_i da \\
&\quad - \int_V (\tilde{\sigma}_{ij,j} + f_i - \rho\ddot{u}_i)\delta u_i dV - \int_{a_\eta} \left[\int_0^t T^{-1}\lambda_{ij}\vartheta_{,i} n_j \mathrm{d}\tau + \eta^*\right]\delta\vartheta \mathrm{d}a \\
&\quad - \int_V \{s + \int_0^t [-T^{-1}\dot{r} - (\lambda_{ij} T^{-1} \vartheta_{,i})_{,j} - \dot{s}^{(i)}]\mathrm{d}\tau\}\delta\vartheta \mathrm{d}V - \int_V \int_0^t \rho_s \ddot{\vartheta} \mathrm{d}\tau \mathrm{d}V\}\delta\vartheta \mathrm{d}V = 0 \\
\tilde{\sigma}_{ij} &= \sigma_{ij} + \sigma_{ij}^T, \quad \sigma_{ij}^T = -(1/2)\left(s + \alpha_{ij}\varepsilon_{ij}\right)\vartheta\delta_{ij} \approx -(1/2)s\vartheta\delta_{ij}
\end{aligned} \tag{43}$$

where σ^T is the effective or equivalent general Maxwell stress which is the external equal axial normal temperature stress. This general Maxwell stress is first introduced and its rationality should be proved by experiments. Obviously σ^T can be neglected for the case of the small strain and small change of temperature. In Eq. (43) it is seen that $\delta\vartheta = \delta_\vartheta\vartheta + \delta_u\vartheta$ is appeared in a whole. Using

$$-T(\lambda_{ij} T^{-1} \vartheta_{,i})_{,j} - T\dot{s}^{(i)} = T\dot{\eta}_{j,j} - T\dot{s}^{(i)} = T\dot{\eta}_{j,j} + \dot{\eta}_i T_{,i} = \left(T\dot{\eta}_i\right)_{,i} = q_{i,i}$$

and the arbitrariness of δui and$\delta\vartheta$, from Eq. (43) we get

$$\begin{aligned} &\tilde{\sigma}_{ij,j} + f_i = \rho \ddot{u}_i; \quad T\left(\dot{s} + \rho_s \ddot{\vartheta}\right) = \dot{r} - q_{i,i} \quad in\ medium \\ &\tilde{\sigma}_{kl} n_l = T_k^*, \quad on \quad a_\sigma; \quad \dot{\eta}_i = -T^{-1} \lambda_{ij} \vartheta_{,i}, \ \dot{\eta}_i n_i = \dot{\eta}_i^*, \ or \quad q_n = q_n^*, \quad on\ a_q \end{aligned} \tag{44}$$

Here $\nabla \cdot \sigma^T$ is the external temperature body force and $n \cdot \sigma T$ is the surface traction.

The above variational principle requests prior that displacements and the temperature satisfy the boundary conditions, so in governing equations the following equations should also be added

$$\mathbf{u} = \mathbf{u}^*, \quad \text{on } a_u; \quad \vartheta = \vartheta^* (or\ T = T^*), \quad \text{on } a_T \tag{45}$$

Eqs. (44) and (45) are the governing equations of the thermo-elasticity derived from the physical variational principle.

3.3. Physical variational principle in thermo-diffusion theory

The electro-chemical Gibbs free energy $g\mu$ and the complement dissipative energy $h\mu$ are expressed in Eq. (33) and the constitutive and evolution equations are expressed in Eq. (34).

Under assumptions that the mechanical displacementu, the temperature ϑ and the chemical potential μ satisfy their own boundary conditionsu=u*, $\vartheta=\vartheta*$and $\mu=\mu*$ onau,aT and aμrespectively. When the variation of temperature is not large the physical variational principle for the thermo-elasto-diffusive problem is

$$
\begin{aligned}
&\delta \Pi_{\mu} = \int_V \delta(g_{\mu} + h_{\mu}) dV + \int_V g_{\mu}^{(T)} \delta u_{k,k} \mathrm{d}V - \delta Q^* - \delta \Phi^* - \delta W = 0 \\
&\delta Q^* = -\int_V \left(\int_0^t T^{-1} \dot{r} d\tau \right) \delta \vartheta \mathrm{d}V + \int_V s^{(a)} \delta \vartheta \mathrm{d}V + \int_{a_q} \eta^* \delta \vartheta \mathrm{d}a \\
&\qquad + \int_V \int_0^t T^{-1} \left(T_{,i} \dot{\eta}_i + \mu_{,i} \dot{\xi}_i \right) \delta \vartheta \mathrm{d}\tau \mathrm{d}V - \int_a \int_0^t T^{-1} \mu \dot{\xi}_i n_i \delta \vartheta \mathrm{d}\tau \mathrm{d}a \\
&\delta \Phi^* = \int_V c^{(a)} \delta \mu \mathrm{d}V + \int_{a_d} \xi^* \delta \mu \mathrm{d}a \\
&\delta W = \int_V (f_k - \rho \ddot{u}_k) \delta u_k dV + \int_{a_\sigma} T_k^* \delta u_k da
\end{aligned}
\tag{46}
$$

In Eqs. (46) f_k, T_k^* $\dot{\eta}^* = \dot{\eta}_i^* n_i$ and $\dot{\xi}^* = \dot{\xi}_i^* n_i$ are given values. In Eq. (46)δQ∗ is related to heat (including the heat produced by the irreversible process in the material), δΦ∗ is related to the diffusion energy. Eq. (46) shows that there is no term in ∫VδhμdV corresponding to the term $-\int_a \int_0^t T^{-1} \mu \dot{\xi}_n \delta \vartheta \mathrm{d}\tau \mathrm{d}a$, , so it should not be included in δQ∗ and $\int_V \int_0^t T^{-1} \mu_{,i} \dot{\xi}_i \delta \vartheta \mathrm{d}\tau \mathrm{d}V - \int_a \int_0^t T^{-1} \mu \dot{\xi}_i n_i \delta \vartheta \mathrm{d}\tau \mathrm{d}a = -\int_V \int_0^t T^{-1} \mu \dot{\xi}_{i,i} \delta \vartheta \mathrm{d}\tau \mathrm{d}V$.

It is noted that we have the following relations

$$
\begin{aligned}
&\int_V \delta g_{\mu} dV = \int_a \sigma_{ij} n_j \delta u_i da - \int_V \sigma_{ij,j} \delta u_i dV - \int_V s \delta \vartheta dV - \int_V c \delta \mu dV \\
&\int_V g_{\mu}^{(T)} \delta u_{k,k} \mathrm{d}V = -(1/2) \int_V \left(s\vartheta + c\mu + \alpha_{ij} \varepsilon_{ij} \vartheta + b_{ij} \varepsilon_{ij} \mu \right) \delta u_{k,k} \mathrm{d}V \\
&\qquad = -(1/2) \int_a \left(s\vartheta + c\mu + \alpha_{ij} \varepsilon_{ij} \vartheta + b_{ij} \varepsilon_{ij} \mu \right) n_k \delta u_k \mathrm{d}V \\
&\qquad + (1/2) \int_V \left(s\vartheta + c\mu + \alpha_{ij} \varepsilon_{ij} \vartheta + b_{ij} \varepsilon_{ij} \mu \right)_{,k} \delta u_k \mathrm{d}V \\
&\delta h_{\mu} = -\int_V \left\{ \delta \vartheta_{,i} \int_0^t \left(\lambda_{ij} T^{-1} T_{,i} + L_{ij} T^{-1} \mu_{,i} \right) d\tau + \delta \mu_{,i} \int_0^t \left(L_{ij} T_{,i} + D_{ij} \mu_{,i} \right) d\tau \right\} dV
\end{aligned}
\tag{47}
$$

The further derivation is fully similar to that in the thermo-elasticity. Combining Eqs. (46) and (47) we get

$$\begin{aligned}
\delta \Pi_\mu = & \int_{a_\sigma} (\tilde{\sigma}_{ij} n_j - T_i^*) \delta u_i da - \int_V (\tilde{\sigma}_{ij,j} + f_i - \rho \ddot{u}_i) \delta u_i dV - \int_V s \delta \vartheta dV - \int_V c \delta \mu dV \\
& - \int_a \left\{ \delta \vartheta n_j \int_0^t \left(\lambda_{ij} T^{-1} T_{,i} + L_{ij} T^{-1} \mu_{,i} \right) d\tau + \delta \mu n_j \int_0^t \left(L_{ij} T_{,i} + D_{ij} \mu_{,i} \right) d\tau \right\} da \\
& + \int_V \left\{ \delta \vartheta \int_0^t \left(\lambda_{ij} T^{-1} T_{,i} + L_{ij} T^{-1} \mu_{,i} \right)_{,j} d\tau + \delta \mu \int_0^t \left(L_{ij} T_{,i} + D_{ij} \mu_{,i} \right)_{,j} d\tau \right\} dV \\
& + \int_V \left(\int_0^t T^{-1} \dot{r} d\tau \right) \delta \vartheta \mathrm{d}V - \int_V s^{(a)} \delta \vartheta \mathrm{d}V - \int_V c^{(a)} \delta \mu \mathrm{d}V \\
& - \int_{a_q} \eta^* \delta \vartheta \mathrm{d}a - \int_{a_d} \xi^* \delta \mu \mathrm{d}a - \int_V \int_0^t \left(T^{-1} T_{,i} \dot{\eta}_i - T^{-1} \mu \dot{\xi}_{i,i} \right) \delta \vartheta \mathrm{d}\tau \mathrm{d}V = 0
\end{aligned} \tag{48}$$

Where

$$\tilde{\sigma}_{ij} = \sigma_{ij} + \sigma_{ij}^{T\mu}, \quad \sigma_{ij}^{T\mu} = -(1/2)\left(s\vartheta + c\mu + \alpha_{ij}\varepsilon_{ij}\vartheta + b_{ij}\varepsilon_{ij}\mu\right)\delta_{ij} \approx -(1/2)(s\vartheta + c\mu) \tag{49}$$

Due to the arbitrariness of δu ,$\delta\vartheta$ and$\delta\mu$, from Eq. (48) we get

$$\tilde{\sigma}_{kl,l} + f_k = \rho \ddot{u}_k, \quad in\ medium; \quad \tilde{\sigma}_{kl} n_l = T_k^*, \quad on\ a_\sigma \tag{50}$$

and

$$\int_V\int_0^t\left(\dot{s}+\rho_s\ddot{\vartheta}\right)d\tau dV=\int_V\int_0^t T^{-1}\left(\dot{r}-q_{j,j}+\mu\dot{\xi}_{j,j}\right)d\tau dV$$
$$\int_V\int_0^t\left(\dot{c}+\rho_c\ddot{\mu}\right)d\tau dV=-\int_V\int_0^t\dot{\xi}_{j,j}d\tau dV \quad \textit{In medium}$$
$$\dot{\eta}_j=-T^{-1}\left(\lambda_{ij}T_{,i}+L_{ij}\mu_{,i}\right),\quad \dot{\eta}_j n_j=\dot{\eta}^*,\quad or\quad q_n=q_n^*\quad on\ a_q$$
$$\dot{\xi}_i=-\left(L_{ij}T_{,i}+D_{ij}\mu_{,i}\right),\quad \dot{\xi}_j n_j=\dot{\xi}^*,\quad or\quad d_n=d_n^*\quad on\ a_d \tag{51}$$

Where $T^{-1}\left(\lambda_{ij}T_{,i}+L_{ij}\mu_{,i}\right)_{,j}-T^{-1}\left(T_{,i}\dot{\eta}_i-\mu\dot{\xi}_{i,i}\right)=-T^{-1}\left(q_{j,j}+\mu\dot{\xi}_{i,i}\right)$ has been used.

The first two formulas in Eq. (51) can be rewritten as

$$T\left(\dot{s}+\rho_s\ddot{\vartheta}\right)=\dot{r}-q_{j,j}+\mu\dot{\xi}_{j,j};\quad \dot{c}+\rho_c\ddot{\mu}=-\dot{\xi}_{j,j}$$
$$T\left(\dot{s}+\rho_s\ddot{\vartheta}\right)+\mu\left(\dot{c}+\rho_c\ddot{\mu}\right)=\dot{r}-q_{j,j};\quad \textit{In mediun} \tag{52}$$

The last equation in Eq. (52) is just the same as that in Eq. (32).

The above variational principle requests prior that the u,ϑ and μ satisfy their own boundary conditions, so in governing equations the following equations should also be added

$$\mathbf{u}=\mathbf{u}^*,\quad \text{on } a_u;\quad \vartheta=\vartheta^*,\quad \text{on } a_T;\quad \mu=\mu^*,\quad \text{on } a_\mu \tag{53}$$

Eqs. (49)-(53) are the governing equations of the generalized thermodiffusion theory. If we neglect the term $\mu\left(\dot{c}+\breve{\dot{c}}^{(a)}\right)$ in Eq. (32), $T\left(\dot{s}+\dot{s}^{(a)}\right)=\dot{r}-q_{i,i}$ is adopted, then we easily get

$$
\begin{aligned}
&T\left(\dot{s}+\rho_s\ddot{\vartheta}\right)=\dot{r}-q_{j,j}, \quad \dot{c}+\rho_c\ddot{\mu}=-\dot{\xi}_{j,j}; \quad In\ medium\\
&\dot{\eta}_j n_j=\dot{\eta}^*, \quad or \quad q_n=q_n^* \quad on\ a_q\\
&\dot{\xi}_j n_j=\dot{\xi}^*, \quad or \quad d_n=d_n^* \quad on\ a_d\ and\ a_q
\end{aligned}
\tag{54}
$$

If we also assume that T,I and μ,j are not dependent each other, then for˙r=0, the Eq. (54) becomes Eq. (36), i.e.

$$
\begin{aligned}
&T\left(\alpha_{ij}\dot{u}_{i,j}+C\dot{\vartheta}/T_0+a\dot{\mu}+\rho_s\ddot{\vartheta}\right)=\lambda_{ij}\vartheta_{,j}\\
&b\dot{\mu}+b_{ij}\dot{u}_{i,j}+a\dot{\vartheta}+\rho_c\ddot{\mu}=D_{ij}\mu_{,j}; \quad In\ medium
\end{aligned}
\tag{55}
$$

3.4. Physical variational principle in electro-magneto-elastic analysis

In this section we discuss the nonlinear electro-magneto-elastic media. Here we extend the theory in previous paper (Kuang, 2011) to the material with the electromagnetic body couple. Because the asymmetric part of the stress is introduced by the electromagnetic body couple, the specific electromagnetic Gibbs free energy gem is taken as

$$
\begin{aligned}
&g_{em}\left(\varepsilon_{kl},E_k,H_k\right)=(1/2)C_{ijkl}\varepsilon_{ji}\varepsilon_{lk}-\left(e^e_{kij}E_k+e^m_{kij}H_k\right)\varepsilon_{ij}-(1/2)\left(\epsilon_{ij}E_iE_j+\mu_{ij}H_iH_j\right)\\
&\quad-(1/2)\left(l^e_{ijkl}E_iE_j+l^m_{ijkl}H_iH_j\right)\varepsilon_{kl}-\left(\epsilon_{km}E_mE_l+\mu_{km}H_mH_l\right)\varepsilon_{kl}\\
&\left(C_{ijkl},l^e_{ijkl},l^m_{ijkl}\right)=\left(C_{jikl},l^e_{jikl},l^m_{jikl}\right)=\left(C_{ijlk},l^e_{ijlk},l^m_{ijlk}\right)=\left(C_{klij},l^e_{klij},l^m_{klij}\right),\ e^e_{kij}=e^e_{kji},\ e^m_{kij}=e^m_{kji}
\end{aligned}
\tag{56a}
$$

where l^{e}_{ijkl} and l^{m}_{ijkl} are the electrostrictive and megnetostrictive constants respectively; ∈and μ may be asymmetric. The corresponding constitutive equations are

$$\begin{aligned}
\sigma_{kl} &= \partial g_{em}/\partial\varepsilon_{kl} = C_{ijkl}\varepsilon_{ij} - e^{e}_{jkl}E_j - e^{m}_{jkl}H_j - (1/2)l^{e}_{ijkl}E_iE_j - (1/2)l^{m}_{ijkl}H_iH_j \\
&\quad - \epsilon_{km}\,E_mE_l - \mu_{km}H_mH_l \\
D_k &= -\partial g_{em}/\partial E_k = \left[\epsilon_{kl} + l^{e}_{ijkl}\varepsilon_{ij} + \left(\epsilon_{ml}\,\varepsilon_{mk} + \epsilon_{mk}\,\varepsilon_{ml}\right)\right]E_l + e^{e}_{kij}\varepsilon_{ij} \approx \epsilon_{kl}\,E_l \\
B_k &= -\partial g_{em}/\partial H_k = \left[\mu_{kl} + l^{m}_{ijkl}\varepsilon_{ij} + \left(\alpha^{m}_{ml}\varepsilon_{mk} + \alpha^{m}_{mk}\varepsilon_{ml}\right)\right]H_l + e^{m}_{kij}\varepsilon_{ij} \approx \mu_{kl}H_l
\end{aligned} \tag{57}$$

Let σs and σa be the symmetric and asymmetric parts of σ

respectively, we have

$$\begin{aligned}
\sigma^{s}_{kl} &= (1/2)(\sigma_{kl} + \sigma_{lk}) = C_{ijkl}\varepsilon_{ji} - e^{e}_{jkl}E_j - (1/2)l^{e}_{ijkl}E_iE_j - e^{m}_{jkl}H_j - (1/2)l^{m}_{ijkl}H_iH_j \\
&\quad - (1/2)(\epsilon_{km}\,E_l + \epsilon_{lm}\,E_k)E_m - (1/2)(\mu_{km}H_l + \mu_{lm}H_k)H_m \\
\sigma^{a}_{kl} &= (1/2)(\sigma_{kl} - \sigma_{lk}) = -(1/2)(\epsilon_{km}\,E_l - \epsilon_{lm}\,E_k)E_m - (1/2)(\mu_{km}H_l - \mu_{lm}H_k)H_m \\
D_kE_l - D_lE_k &= P_kE_l - P_lE_k \approx (\epsilon_{km}\,E_l - \epsilon_{lm}\,E_k)E_m \\
B_kH_l - B_lH_k &= \mu_0(M_kH_l - M_lH_k) \approx (\mu_{km}H_l - \mu_{lm}H_k)H_m
\end{aligned} \tag{58}$$

where D=∈0E+P,B=μ0(H+M) have been used, P and M are the polarization density and magnetization density, ∈0 and μ0 are the dielectric constant and magnetic permeability in vacuum respectively. The terms containing ε in D and B in Eq. (58) have been neglected. In the usual electromagnetic theory the electromagnetic body couple isP×E+μ0M×H . From Eq. (58) it is seen that $2\sigma^{a}_{kl} + (D_kE_l - D_lE_k) + (B_kH_l - B_lH_k) = 0$ or the electromagnetic body couple is balanced by the moment produced by the asymmetric stresses.

Using Eq. (57), Eq. (56a) can be reduced to

$$g_{em} = (1/2)C_{ijkl}\varepsilon_{ji}\varepsilon_{lk} + g^{em},\ g^{em} = -(1/2)(D_k E_k + B_k H_k + \Delta_{kl}\varepsilon_{lk}) \approx -(1/2)(D_k E_k + B_k H_k)$$
$$\Delta_{kl} = e^{e}_{mkl}E_m + e^{m}_{mkl}H_m = -e^{e}_{mkl}\varphi_{,m} - e^{m}_{mkl}\psi_{,m}$$

(56b)

Because the value of the term Δ:ε is much less than that of other terms, it can be neglected. In the nonlinear electro-magneto-elastic analysis the medium and its environment should be considered together as shown in Fig. 1 (Kuang, 2011a, 2011b), because the electromagnetic field exists in all space. Under the assumption that $\mathbf{u},\varphi,\psi,\mathbf{u}^{env},\varphi^{env},\psi^{env}$ satisfy their boundary conditions on their own boundaries au,aφ,aψ,aenvu,aenvφ,aenvψ and the continuity conditions on the interfaceaint. The Physical variational principle in the nonlinear electro-magneto-elastic analysis is

$$\delta\Pi = \delta\Pi_1 + \delta\Pi_2 - \delta W^{*int} = 0$$
$$\delta\Pi_1 = \int_V \delta g_{em} dV + \int_V g^{em}\delta u_{i,i} dV - \delta W^*$$
$$\delta\Pi_2 = \int_{V^{env}} \delta g^{env}_{em} dV + \int_{V^{env}} g^{em\,env}\delta u^{env}_{i,i} dV - \delta W^{*env}$$
$$\delta W^* = \int_V (f_k - \rho\ddot{u}_k)\delta u_k dV - \int_V \rho_e \delta\varphi dV + \int_{a_\sigma} T^*_k \delta u_k da - \int_{a_q} \sigma^* \delta\varphi da + \int_{a_\mu} B^*_i n_i \delta\psi da$$
$$\delta W^{*int} = \int_{a^{int}} T^{*int}_k \delta u_k da - \int_{a^{int}} \sigma^{*int}\delta\varphi da + \int_{a^{int}_\mu} B^{*int}_i n_i \delta\psi^{env} da$$
$$g^{env}_{em},\ g^{em\,env},\ \delta W^{*env}\ \text{are similar and omited here}$$

(59)

where the superscript "env" means the variable in environment, "int" means the variable on the interface, $f_k, T^*_k, \sigma^*, B_n^* = B_i^* n_i;\ T^{*env}_k,;\ \sigma^{*env}, B_n^{*env}; T^{*int}_k, \sigma^{*int}, B_n^{*\ \ *int}$ are the given values on the corresponding surfaces. Eq. (59) is an alternative form of Eq. (39) and the electromagnetic force is directly enclosed in the formula (Kuang, 2008a, 2009a).

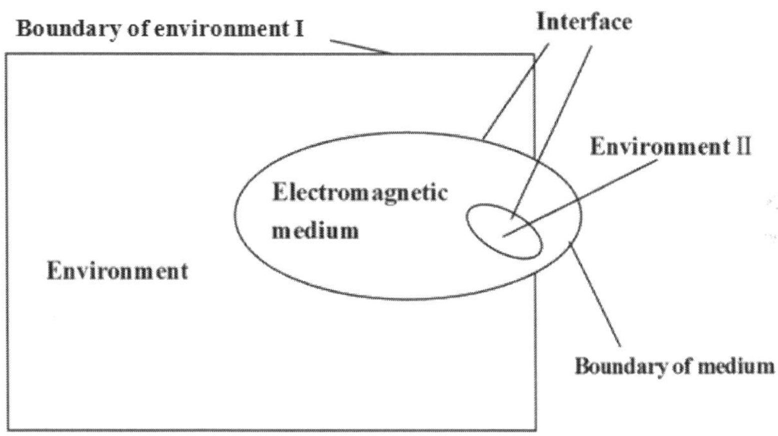

Figure 1. Electromagnetic medium and its environment

As shown in previous paper (Kuang, 2011a, 2011b) and in section 3.1. the variations of φ,ψ,E,H

will be distinguished into local and migratory variations, i.e.

$$
\begin{aligned}
&\delta(\varphi,\psi,E_i,H_i)=\delta_{(\varphi,\psi,E_i,H_i)}(\varphi,\psi,E_i,H_i)+\delta_u(\varphi,\psi,E_i,H_i)\\
&\delta_u(\varphi,\psi)=(\varphi,\psi)_{,p}\delta u_p=-(E_p,H_p)\delta u_p\\
&\delta_u(E_i,H_i)=-\delta_u(\varphi,\psi)_{,i}=-(\varphi,\psi)_{,ip}\delta u_p=(E_{p,i},H_{p,i})\delta u_p=(E_{i,p},H_{i,p})E_{i,p}\delta u_p
\end{aligned}
\tag{60}
$$

Noting that in Eq. (59) we have

$$\int_V \delta g_{em} dV + \int_V g^{em} \delta u_{k,k} dV = \int_V \sigma_{ij} \delta\varepsilon_{ij} dV - \int_V D_i \delta E_i dV - \int_V B_i \delta H_i dV$$
$$-(1/2)\int_V (D_k E_k + B_k H_k)\delta u_{j,j} dV = \int_a \left[\sigma_{ij} - (1/2)(D_k E_k + B_k H_k)\delta_{ij}\right] n_j \delta u_i \mathrm{d}a$$
$$-\int_V \left[\sigma_{ij} - (1/2)(D_k E_k + B_k H_k)\delta_{ij}\right]_{,j} \delta u_i \mathrm{d}V + \int_a D_i n_i \delta_\varphi \varphi \mathrm{d}a - \int_V D_{i,i} \delta_\varphi \varphi \mathrm{d}V$$
$$-\int_V D_i E_{p,i} \delta u_p \mathrm{d}V + \int_a B_i n_i \delta_\psi \psi \mathrm{d}a - \int_V B_{i,i} \delta_\psi \psi \mathrm{d}V - \int_V B_i H_{p,i} \delta u_p \mathrm{d}V$$

SoδΠ1 in Eq. (59) is reduced to

$$\delta\Pi_1 = \delta\Pi_1' + \delta\Pi_1''$$
$$\delta\Pi_1' = \int_{a_\sigma} \left(\sigma_{kj} n_j - T_k^*\right)\delta u_k \mathrm{d}a + \int_{a^{\mathrm{int}}} \sigma_{kj} n_j \delta u_k \mathrm{d}a$$
$$-\int_V \left(\sigma_{kj,j} + f_k - \rho \ddot{u}_k\right)\delta u_k \mathrm{d}V - \int_V \left(D_{i,i} - \rho_e\right)\delta_\varphi \varphi \mathrm{d}V + \int_{a_D} \left(D_i n_i + \sigma^*\right)\delta_\varphi \varphi \mathrm{d}a$$
$$+\int_{a^{\mathrm{int}}} D_i n_i \delta_\varphi \varphi \mathrm{d}a - \int_V B_{i,i} \delta_\psi \psi \mathrm{d}V + \int_{a_\mu} \left(B_i - B_i^*\right) n_i \delta_\psi \psi \mathrm{d}a + \int_{a^{\mathrm{int}}} B_i n_i \delta_\psi \psi \mathrm{d}a$$
$$\delta\Pi_1'' = (1/2)\int_{a_\sigma} \left(D_m \varphi_{,m} + B_m \psi_{,m}\right) n_k \delta u_k \mathrm{d}a + (1/2)\int_{a^{\mathrm{int}}} \left(D_m \varphi_{,m} + B_m \psi_{,m}\right) n_k \delta u_k \mathrm{d}a$$
$$-(1/2)\int_V \left(D_m \varphi_{,m} + B_m \psi_{,m}\right)_{,k} \delta u_k \mathrm{d}V - \int_V D_i E_{p,i} \delta u_p \mathrm{d}V - \int_V \rho_e E_p \delta u_p dV$$
$$-\int_{a_D} \sigma^* E_p \delta u_p da - \int_V B_i H_{p,i} \delta u_p \mathrm{d}V + \int_{a_\mu^{env}} B_i^* n_i H_p \delta u_p da \quad (61)$$

where δΠ'1 is the part of δΠ1 due to the local variations ofu,φ,ψ ; δΠ″1 is the part of δΠ1 due to the migratory variations ofφ,ψ . Substituting the following identity

$$-\int_V D_i E_{p,i} \delta u_p \mathrm{d}V - \int_V \rho_e E_p \delta u_p dV - \int_{a_D} \sigma^* E_p \delta u_p da - \int_V B_i H_{p,i} \delta u_p \mathrm{d}V$$
$$+\int_{a_\mu} B_i^* n_i H_p \delta u_p da = \int_{a_D} (D_i n_i + \sigma^*)\delta_u \varphi da + \int_{a^{\mathrm{int}}} D_i n_i \delta_u \varphi da + \int_a \left(D_i E_p\right) n_i \delta u_p da$$
$$-\int_V \left(D_i E_p\right)_{,i} \delta u_p dV - \int_V (D_{i,i} - \rho_e)\delta_u \varphi dV + \int_{a_\mu} \left(B_i - B_i^*\right) n_i \delta_u \psi da$$
$$+\int_{a^{\mathrm{int}}} B_i n_i \delta_u \psi da + \int_a \left(B_i H_p\right) n_i \delta u_p da - \int_V \left(B_i H_p\right)_{,i} \delta u_p dV - \int_V B_{i,i} \delta_u \psi dV \quad (62)$$

into δΠ″1 in Eq. (61) we get

$$\begin{aligned}\delta\Pi_1'' = &-\int_V (D_{i,i} - \rho_e)\delta_u\varphi dV + \int_{a_D} (D_i n_i + \sigma^*)\delta_u\varphi da + \int_{a^{\text{int}}} D_i n_i \delta_u \varphi da \\ &- \int_V B_{i,i}\delta_u\psi dV + \int_{a_\mu} \left(B_i - B_i^*\right) n_i \delta_u \psi da + \int_{a^{\text{int}}} B_i n_i \delta_u \psi da \\ &+ \int_{a_\sigma} \sigma_{ik}^M n_i \delta u_k da + \int_{a^{\text{int}}} \sigma_{ik}^M n_i \delta u_k da - \int_V \sigma_{ik,i}^M \delta u_k dV\end{aligned} \tag{63}$$

where σM is the Maxwell stress:

$$\sigma_{ik}^M = D_i E_k + B_i H_k - (1/2)(D_n E_n + B_n H_n)\delta_{ik} \tag{64}$$

Substituting Eq. (63) into Eq. (61) we get

$$\begin{aligned}\delta\Pi_1 = &\int_{a_\sigma} (\tilde\sigma_{ij} n_i - T_j^*)\delta u_j \mathrm{d}a + \int_{a^{\text{int}}} \tilde\sigma_{ij} n_i \delta u_j \mathrm{d}a - \int_V (\tilde\sigma_{ij,i} + f_j - \rho\ddot u_j)\delta u_j \mathrm{d}V \\ &+ \int_{a_D} (D_{i,i} + \sigma^*)\delta\varphi \mathrm{d}a + \int_{a^{\text{int}}} D_i n_i \delta\varphi \mathrm{d}a - \int_V (D_{i,i} - \rho_e)\delta\varphi dV \\ &+ \int_{a_\mu} \left(B_i - B_i^*\right) n_i \delta\psi da + \int_{a^{\text{int}}} B_i n_i \delta\psi da - \int_V B_{i,i}\delta\psi dV \\ \tilde\sigma_{kl} = &\sigma_{kl} + \sigma_{kl}^M = C_{ijkl}\varepsilon_{ij} - e_{jkl}^e E_j - e_{jkl}^m H_j - (1/2) l_{ijkl}^e E_i E_j - (1/2) l_{ijkl}^m H_i H_j \\ &-(1/2)(D_m E_m + B_m H_m)\delta_{kl}\end{aligned} \tag{65}$$

where ˜σ is the pseudo total stress (Jiang and Kuang, 2004), which is not the true stress in electromagnetic media. From the expression of ˜σ it is known that ˜σ is symmetric though σ and σM are asymmetric. Due to the arbitrariness of δui,δφ andδψ, from Eq. (65) we get

$$\tilde{\sigma}_{jk,j} + f_k = \rho \ddot{u}_k, \quad D_{i,i} = \rho_e, \quad B_{i,i} = 0, \quad \text{in } V$$
$$\tilde{\sigma}_{jk} n_j = T_k^*, \quad on\ a_\sigma; \quad D_i n_i = -\sigma^*, \quad on\ a_D; \quad \left(B_i - B_i^*\right) n_i = 0, \quad on\ a_\mu$$
$$\delta \Pi_1 = \int_{a^{\text{int}}} \tilde{\sigma}_{ij} n_i \delta u_j \mathrm{d}a + \int_{a^{\text{int}}} D_i n_i \delta \varphi da + \int_{a^{\text{int}}} B_i n_i \delta \psi da \tag{66}$$

For the environment we have the similar formula:

$$\tilde{\sigma}_{ij,i}^{env} + f_j^{env} = \rho^{env} \ddot{u}_j^{env}, \quad D_{i,i}^{env} = \rho_e^{env}, \quad B_{i,i} = 0, \quad \text{in } V^{env}$$
$$\tilde{\sigma}_{ij}^{env} n_i^{env} = T_j^{*env}, \ \text{on } a_\sigma^{env}; \ D_i^{env} n_i^{env} = -\sigma^{*env}, \ on\ a_D^{env}; \ B_i^{env} n_i^{env} = B_i^{*env} n_i^{env}, \ on\ a_\mu^{env}$$
$$\delta \Pi_2 = \int_{a^{\text{int}}} \tilde{\sigma}_{ij}^{env} n_i^{env} \delta u_j^{env} da + \int_{a^{\text{int}}} D_i^{env} n_i^{env} \delta \varphi^{env} da + \int_{a^{\text{int}}} B_i^{env} n_i^{env} \delta \psi^{env} da$$
$$\tilde{\sigma}_{jk}^{env} = \sigma_{jk}^{env} + \sigma_{jk}^{M\,env} \tag{67}$$

Using $n_i = -n_i^{env}$, $u_i = u_i^{env}$, $\varphi = \varphi^{env}$, $\psi = \psi^{env}$ on the boundary surface we get

$$(\tilde{\sigma}_{ij} - \tilde{\sigma}_{ij}^{env}) n_i = T_j^{*\text{int}}, \ (D_i - D_i^{env}) n_i = -\sigma^{*\text{int}}, \ (B_i - B_i^{env}) n_i = B_i^{*\text{int}} n_i, \ \text{on } a^{\text{int}} \tag{68}$$

The above variational principle requests prior that the displacements, the electric potential and the magnetic potential satisfy their own boundary conditions and the continuity conditions on the interface, so the following equations should also be added to governing equations

$$u_i = u_i^*, \ \text{on } a_u; \ \ \varphi = \varphi^*, \ \text{on } a_\varphi; \ \ \psi = \psi^*, \ \text{on } a_\mu$$
$$u_i^{env} = u_i^{*env}, \ \text{on } a_u^{env}; \ \ \varphi^{env} = \varphi^{*env}; \ on \ a_\varphi^{env}; \ \ \psi^{env} = \psi^{*env}, \ on \ a_\mu^{env}$$
$$u_i = u_i^{env}, \ \varphi = \varphi^{env}, \ \psi = \psi^{env}; \ on \ a^{\text{int}} \tag{69}$$

Eqs. (66)—(69) are the governing equations. It is obvious that the above physical variational principle is easy to extend to other materials.

3.5. Materials with static magnetoelectric coupling effect

In this section we discuss the electro-magneto-elastic media with static magnetoelectric coupling effect shortly. For these materials the constitutive equations are

$$\begin{aligned} \sigma_{kl} &= C_{ijkl}\varepsilon_{ij} - e^e_{jkl}E_j - e^m_{jkl}H_j - (1/2)l^e_{ijkl}E_iE_j - (1/2)l^m_{ijkl}H_iH_j \\ &\quad - \epsilon_{km}E_mE_l - \mu_{km}H_mH_l - \beta_{km}H_mE_l - \beta_{km}E_mH_l \\ D_k &= \left[\epsilon_{kl} + l^e_{ijkl}\varepsilon_{ij} + (\epsilon_{ml}\varepsilon_{mk} + \epsilon_{mk}\varepsilon_{ml})\right]E_l + e^e_{kij}\varepsilon_{ij} + \beta_{kl}H_l \\ B_k &= \left[\mu_{kl} + l^m_{ijkl}\varepsilon_{ij} + 2\left(\alpha^m_{ml}\varepsilon_{mk} + \alpha^m_{mk}\varepsilon_{ml}\right)\right]H_l + e^m_{kij}\varepsilon_{ij} + \beta_{kl}E_l \end{aligned} \tag{70}$$

where $\beta ij=\beta ji$ is the static magnetoelectric coupling coefficient. The electromagnetic body couple is still balanced by the asymmetric stress, i.e.

$$D_k E_l - D_l E_k + B_k H_l - B_l H_k = \left[(\epsilon_{km} E_l - \epsilon_{lm} E_k) E_m + (\mu_{km} H_l - \mu_{lm} H_k) H_m \right] + \left[(\beta_{km} E_l - \beta_{lm} E_k) H_m + (\beta_{km} H_l - \beta_{lm} H_k) E_m \right] = -2\sigma_{kl}^{a}$$

In this case though the constitutive equations are changed, but the electromagnetic Gibbs free energy ge in Eq. (56b), governing equations (66)—(69) and the Maxwell stress (64) are still tenable.

4. CONCLUSIONS

In this chapter some advances of thermodynamics in continuum mechanics are introduced. We advocate that the first law of the thermodynamics includes two contents: one is the energy conservation and the other is the physical variational principle which is substantially the momentum equation. For the conservative system the complete governing equations can be obtained by using this theory and the classical thermodynamics. For the non-conservative system the complete governing equations can also be obtained by using this theory and the irreversible thermodynamics when the system is only slightly deviated from the equilibrium state. Because the physical variational principle is tensely connected with the energy conservation law, so we write down the energy expressions, we get the physical variational principle immediately and do not need to seek the variational functional as that in usual mathematical methods.

In this chapter we also advocate that the accelerative variation of temperature needs extra heat and propose the general inertial entropy theory. From this theory the temperature wave and the diffusion wave with finite propagation velocities are easily obtained. It is found that the coupling effect in elastic and temperature waves attenuates the temperature wave, but enhances the elastic wave. So the theory with two parameters by introducing the viscous effect in this problem may be more appropriate.

Some explanation examples for the physical variational principle and the inertial entropy theory are also introduced in this chapter, which may indirectly prove the rationality of these theories. These theories should still be proved by experiments.

REFERENCES

1. R. M. Christensen, 2003 "Theory of Viscoelasticity", Academic Press, New York.

2. S. R. De Groet, 1952 "Thermodynamics of Irreversible Processes", North-Holland Publishing Company,

3. A. E. Green, K. A. Lindsay, 1972 "Thermoelasticity", Journal of Elasticity, 2 17 .

4. I. Gyarmati, 1970 "Non-equilibrium thermodynamics, Field theory and variational principles", Berlin, Heidelberg, New York, Springer-Verlag.

5. Z. Kuang, B. , 1999 "Some remarks on thermodynamic theory of viscous-elasto-plastic media", in IUTAM symposium on rheology of bodies with defects, 8799 , Ed. By Wang, R., Kluwer Academic Publishers.

6. Z. Kuang, B. , 2002 "Nonlinear continuum mechanics", Shanghai Jiaotong University Press, Shanghai. (in Chinese)

7. Z. Kuang, B. , 2007 "Some problems in electrostrictive and magnetostrictive materials", Acta Mechanica Solida Sinica, 20 219217 .

8. Z. Kuang, B., 2008a "Some variational principles in elastic dielctric and elastic magnetic materials", European Journal of Mechanics- A/Solids, 27 504514 .

9. Z. Kuang, B. , 2008b "Some variational principles in electroelastic media under finite deformation", Science in China, Series G, 51 13901402 .

10. Z. Kuang, B. , 2009a "Internal energy variational principles and governing equations in electroelastic analysis", International journal of solids and structures, 46 902911 .

11. Z. Kuang, B. , 2009b "Variational principles for generalized dynamical theory of thermopiezoelectricity", Acta Mechanica, 203 111 .

12. Kuang Z.-B. 2010 "Variational principles for generalized thermodiffusion theory in pyroelectricity", Acta Mechanica, 214 275289 .

13. Z. Kuang, B. , 2011a "Physical variational principle and thin plate theory in electro-magneto-elastic analysis", International journal of solids and structures, 48 317325 .

14. Z. Kuang, B. , 2011b "Theory of Electroelasticity", Shanghai Jiaotong University Press, Shanghai (in Chinese).

15. D. Jou, J. Casas-Vzquez, G. Lebon, 2001 „Extended irreversible thermodynamics (third, revised and enlarged edition)", Springer-Verlag, Berlin, Heidelberg, New York.

16. H. W. Lord, Y. Shulman, 1967 "A generalized dynamical theory of thermoelasticity", Journal of the Mechanics and Physics of Solids, 15 299309 .

17. H. H. Sherief, F. A. Hamza, H. Saleh, 2004, Saleh, H. A. 2004, "The theory of generalized thermoelastic diffusion", International Journal of engineering science, 42: 591—608.

18. Z. X. Wang, 1955 Thermodynamics, Higher Education Press, Beijing. (in Chinese).

19. X. G. Yuan, Z. Kuang-B, 2008 Waves in pyroelectrics[J], Journal of thermal stress, 31 11901211 .

20. X. G. Yuan, Z. Kuang-B, 2010 "The inhomogeneous waves in pyroelectrics" [J], Journal of thermal stress, 33 172186 .

21. A. Yunus, Michael. A. Çengel, Boles, 2011 Thermodynamics : an engineering approach, 7th ed, McGraw-Hill, New York.

CHAPTER 4

Nonlinear vibration behavior of graphene resonators and their applications in sensitive mass detection

Mai Duc Dai[1], Chang-Wan Kim[1*] and Kilho Eom [2,3*]

1Department of Mechanical Engineering, Konkuk University, Seoul 143-701,Republic of Korea.

2Department of Biomedical Engineering, Yonsei University,Wonju 220-740, Republic of Korea.

3Institute for Molecular Sciences, Seoul, 120-749, Republic of Korea.

ABSTRACT

Graphene has received significant attention due to its excellent mechanical properties, which has resulted in the emergence of graphene-based nano-electro-mechanical system such as nanoresonators. The nonlinear vibration of a graphene resonator and its application to mass sensing (based on nonlinear oscillation) have been poorly studied, although a graphene resonator is able to easily reach the nonlinear vibration. In this work, we have studied the nonlinear vibration of a graphene resonator driven by a geometric nonlinear effect due to an edge-clamped boundary condition using a continuum elastic model such as a plate model. We have shown that an in-plane tension can play a role in modulating the nonlinearity of a resonance for a graphene. It has been found that the detection sensitivity of a graphene resonator can be improved by using nonlinear vibration induced by an actuation force-driven geometric nonlinear effect. It is also shown that an in-plane tension can control the detection sensitivity of a graphene resonator that operates both harmonic and nonlinear oscillation regimes. Our study suggests the design principles of a graphene resonator as a mass sensor for developing a novel detection scheme using graphene-based nonlinear oscillators.

KEYWORDS

Graphene resonator; Mass sensing Nonlinear; oscillation NEMS

1. BACKGROUND

Graphene has recently been attracting the scientific community due to its excellent electrical [1–3] and/or mechanical properties [4–8]; these remarkable properties have enabled the exploitation of graphene for the development of nano-electro-mechanical system (NEMS) such as nanoresonators [9, 10]. Specifically, since a pioneering work by researchers at Cornell [11], graphene has recently been extensively taken into account for designing nanoresonators that can exhibit high-frequency dynamic range [11–13] with favorable high Q factors [13–17]. The high-frequency dynamics of graphene is attributed to its excellent mechanical properties such as Young's modulus of approximately 1 TPa [4–8, 18]; it is noted that a resonant frequency is linearly proportional to the square root of Young's modulus when a device operates in harmonic oscillation [9, 10]. Until recently, most research works [11–15] (except a work by Eichler et al. [17]) have focused on the harmonic oscillation of a graphene resonator. However, the nonlinear vibration of a graphene resonator has not been well studied yet, albeit a recent study [17] reports an experimental observation of the nonlinear vibration of a graphene resonator. The nonlinear elastic deformation of a graphene is ubiquitous due to the fact that a monolayer graphene is an atomically thin sheet so that the out-of-plane deflection of a graphene is much larger than its thickness [19], which indicates that a graphene can easily undergo a nonlinear elastic deflection. Moreover, as discussed in our previous study [9, 20, 21], the nonlinear vibration is a useful route to the development of novel sensitive detection scheme based on nanoresonators made of nanomaterials such as carbon nanotubes.

To gain a detailed insight into the underlying mechanism of the vibration of a graphene resonator, an atomistic simulation such as molecular dynamics (MD) simulation has been widely utilized. For instance, Park and coworkers [22, 23] have studied various effects such as edge effect and/or internal friction effect on the vibrational behavior of graphene resonators using MD simulation. Furthermore, Park and coworkers [24] have investigated the energy dissipation mechanism of vibrating polycrystalline graphenes fabricated from chemical vapor deposition method by using MD simulation. Despite the ability of MD simulation to provide detailed characteristics of the vibrational behavior of graphene resonators, MD simulation is computationally restricted to studying the vibrational behavior of a graphene resonator whose length scale is <10 nm (e.g., see refs. [22,

24]). On the other hand, most experimental studies have considered a graphene resonator whose length scale is >1 μm (e.g., see refs. [11–15]). This clearly indicates that a current atomistic simulation is unable to be utilized to analyze an experimentally observed vibrational behavior of a graphene resonator whose length scale is in the order of micrometer.

The computational limitation of atomistic simulations in depicting the underpinning principles of experimentally observed mechanics of graphene resonators has led researchers [19, 25–27] to consider a continuum elastic model, particularly a plate model, for unveiling the vibrational characteristics of a graphene resonator. In order for a continuum elastic model to dictate the atomistic feature of the mechanics of a graphene, the elastic constants of a continuum elastic model (e.g., plate model) for a graphene have to be determined from an atomistic simulation such as MD simulation as it was taken into account for deciding the elastic constants of atomic structures (e.g., lattice) [7, 8]. Recently, a plate model with its elastic constants obtained from MD simulation has allowed revealing the mechanisms of the mechanics of a graphene. More remarkably, in a recent study by Isacsson and coworkers [19], a plate model has been utilized for studying the vibrational behavior of a graphene resonator; it is shown that the vibrational behavior of a graphene resonator predicted from a plate model, whose elastic constants were determined from atomistic model, is consistent with an experimentally observed vibration of a graphene resonator. However, a recent study by Isacsson et al. [19] has only concentrated on the harmonic oscillation of a graphene resonator, even though a graphene resonator can easily reach the nonlinear vibration regime. To the best of our knowledge, despite recent studies [28, 29] theoretically reporting the nonlinear vibration of a graphene resonator, the nonlinear oscillation of a graphene resonator (particularly, nonlinearity tuning), as well as atomic mass detection using graphene-based nonlinear oscillators, has not been well studied based on a continuum elastic model and/or MD simulation.

In this work, we have studied the nonlinear vibration of a graphene resonator using a continuum elastic model, i.e., plate model. We have found that nonlinear oscillation is a useful avenue for improving the detection sensitivity of a graphene resonator and that the detection sensitivity of a graphene-based nonlinear oscillator is governed by both the actuation force (which determines the nonlinearity of vibration) and the size of a graphene resonator. It is shown that the nonlinearity of vibration for a graphene resonator can be tuned by an in-plane tension and that such in-plane tension can modulate the detection sensitivity of a graphene resonator that operates in both harmonic and nonlinear oscillations. In particular, an in-plane tension improves the dynamic frequency range and detection sensitivity of a graphene resonator that operated in harmonic oscillation, while an in-plane tension deteriorates the dynamic frequency range and sensing performance of a graphene-based nonlinear oscillator. Our study sheds light on a continuum elastic model for gaining insight into not only the underlying mechanisms of nonlinear vibration-based enhancement of the dynamic frequencies and sensing performance of a graphene resonator, but also the role of an in-plane tension in modulating the nonlinearity of a graphene resonator.

2. METHODS

2.1 Theory and model

Graphene can be modeled as a plate whose mechanical deformation is attributed to the strain energy composed of bending energy U_B and stretching energy U_S represented in the form [30]

$$U_B = \frac{1}{2}\int_{\Omega} K\left[\nabla^2 w(x,y,t)\right]^2 dxdy \tag{1.a}$$

$$\begin{aligned} U_s = &\frac{1}{2}\int_{\Omega} \frac{E_S h}{4(1-\nu^2)} \\ &\times \left[\{\partial_x w(x,y,t)\}^2 + \{\partial_y w(x,y,t)\}^2\right]^2 dxdy \\ &+\frac{1}{2}\int_{\Omega} N_0 \cdot [\nabla w(x,y,t)] dxdy \end{aligned} \tag{1.b}$$

where κ $E_S h$, and ν represent the bending rigidity, axial stretching modulus, thickness, and Poisson's ratio of a graphene, respectively, $w(x\ y\ t)$ indicates the out-of-plane deflection of a graphene, x and y are the coordinates along the in-plane direction of a graphene, respectively, $\mathbf{N}_0$ is a constant axial tension (due to pre-strain) applied to a graphene, and the symbol Ω in an integrand indicates the surface integral. The strain energy can be related to a potential field prescribed to the atomic structure of a graphene, as has been elucidated in Cauchy-Born model [31–34], such as

$$\sum_{i=1}^{N} U_i^{\text{atom}}(\mathbf{r}) = U_B + U_S \tag{2}$$

Here, U_i^{atom} is a potential field prescribed to an i-th carbon atom for a graphene, $\mathbf{R}$ is the atomic coordinates of a graphene, and N is the total number of carbon atoms for a graphene. Here, it should be noted that when the elastic constants of a graphene (i.e. κ and E_S) are determined from Equation 2, an axial tension $\mathbf{N}_0$ is assumed to be zero (i.e., pre-stress is not applied to a graphene). As described in a literature [25], the force field parameters provide the elastic constants of a graphene such as κ = 1.5 eV and $E_S h$ = 2,000 eV/nm^2.

In order to obtain the equation of motion, we need to know the kinetic energy T for a vibrating graphene. The kinetic energy T can be written as

$$T = \frac{1}{2}\int_{\Omega} \rho_0 [\partial_t w(x, y, t)]^2 dxdy \tag{3}$$

where ρ_0 is the mass density of a graphene; the mass density ρ_0 can be straightforwardly determined from a relation of $\rho_0 = Nm_C/S$, where m_C is the atomic mass of a carbon atom, and S is the surface area of a graphene. The equation of motion for a vibrant graphene can be obtained from the minimization of a Hamiltonian H defined as $H = U + T - W$, where W is the work done by an external force field such as actuation force. The variation of the Hamiltonian, δH, can be obtained as [30, 35]

$$\delta H = \left[\rho_0 \partial_t^2 w + \kappa \nabla^4 w - \frac{E_S}{2(1-\nu)}\left\{(\partial_x w)^2 + (\partial_y w)^2\right\} \times \nabla^2 w - \frac{E_S}{1+\nu}\partial_x w \partial_y w \partial_{xy} w - N_0^{\alpha} \partial_{\alpha}^2 w - f\right]\delta w = 0 \tag{4}$$

where f is an actuation force per unit area for a graphene, a symbol δ indicates a variation, δw is a virtual out-of-plane deflection of a graphene, a Greek index indicates the coordinates, i.e., $\alpha = x$ (for α =

1) or y (for $\alpha = 2$), and a repeated Greek symbol represents Einstein's summation rule. The equation of motion is therefore given by

$$\rho_0 \partial_t^2 w + \kappa \nabla^4 w - \frac{E_S}{2(1-\nu)} \left\{ (\partial_x w)^2 + (\partial_y w)^2 \right\} \nabla^2 w - \frac{E_S}{1+\nu} \partial_x w \partial_y w \partial_{xy} w - N_0^\alpha \partial_\alpha^2 w = f \quad (5)$$

Here, it should be noted that in-plane displacements are ignored in the governing equation given by Equation 5 since in-plane displacements are small in comparison with out-of-plane displacement $w(x\ t)$. In this work, for theoretical convenience, we assume that axial force $\mathbf{N}_0$ is the biaxial loading represented in the form of $\mathbf{N}_0 = N_0(\mathbf{E}_x + \mathbf{E}_y)$, where $\mathbf{E}_x$ and $\mathbf{E}_y$ indicate the directional unit vectors in the x and y directions, respectively. Furthermore, the actuation force f is assumed to be in the form of $f = f_0 \cos \Omega t$, where f_0 is the amplitude of an actuation force, and Ω is a driving frequency. For solving the equation of motion given by Equation 5, we assume that the out-of-plane displacement, i.e., $w(x\ y\ t)$, can be decomposed in the following form [9, 36–38]

$$w(x, y, t) = z(t) \cdot \psi(x, y) \quad (6)$$

Here, $z(t)$ indicates a time-dependent amplitude, and $\psi(x\ y)$ represents the deflection eigenmode for a vibrant graphene. In our work, we presume that a monolayer graphene exhibits a rectangular shape and that all edges of a graphene are clamped. The deflection eigenmode that satisfies the clamped boundary conditions is represented in the form

$$\psi(x,y) = \frac{2}{3}\left[1 - \cos\left(\frac{2\pi x}{a}\right)\right]\left[1 - \cos\left(\frac{2\pi y}{b}\right)\right] \tag{7}$$

where a and b are the lengths of the graphene edges, respectively (see Figure 1). By substituting Equation 6 into Equation 5 followed by integration by parts, the equation of motion represented in Equation 5 becomes the Duffing equation [39–41] as follows

$$\mu\partial_t^2 z(t) + \alpha z(t) + \lambda[z(t)]^3 = p_0 \cos\Omega t \tag{8}$$

where the parameters μ α λ, and p_0 are given by

$$\mu = \rho\int_0^a\int_0^b \psi^2(x,y)dxdy = \rho ab \tag{9.a}$$

$$\alpha = \int_0^a\int_0^b \kappa\left[\nabla^2\psi(x,y)\right]^2 dxdy - N_0\int_0^a\int_0^b [\nabla\psi(x,y)]^2 dxdy = \frac{16\pi^4\kappa}{9a^3b^3}\left[3(a^4+b^4)+2a^2b^2\right] + \frac{4\pi^2 N_0}{3ab}(a^2+b^2) \tag{9.b}$$

$$\lambda = \int_0^a \int_0^b \frac{Eh}{2(1-\nu^2)} \psi(x,y)\left(\partial_x \psi(x,y)\right)^2 + \nu\left(\partial_y \psi(x,y)\right)^2]\partial_x^2 \psi(x,y)dxdy$$
$$- \int_0^a \int_0^b \frac{Eh}{2(1-\nu^2)} \psi(x,y)\left[\nu\left(\partial_x \psi(x,y)\right)^2 + \left(\partial_y \psi(x,y)\right)^2\right] \times \partial_y^2 \psi(x,y)dxdy$$
$$- \int_0^a \int_0^b \frac{Eh}{1+\nu} \psi(x,y)\partial_x \psi(x,y)\partial_y \psi(x,y)\partial_{xy}\psi(x,y)dxdy$$
$$= \frac{Ehab}{81(1-\nu^2)}\left[70\left\{\left(\frac{\pi}{a}\right)^4 + \left(\frac{\pi}{b}\right)^4\right\} + \frac{100\pi^4}{a^2b^2}(4\nu - 1)\right] \quad (9.c)$$

$$p_0 = f_0 \int_0^a \int_0^b \psi^2(x,y)dxdy = f_0 ab \quad (9.d)$$

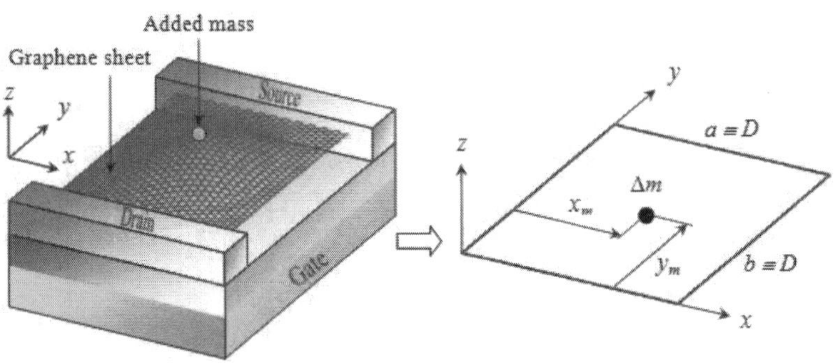

Figure 1 Schematic Illustration Of A Suspended Graphene Resonator Onto Whose Surface Atomic Mass Is Adsorbed.

The vibrational behavior of a graphene resonator can be numerically described by solving the Duffing equation given by Equation 8.

For a case in which atoms are adsorbed onto a graphene resonator, as shown in Figure 1, we have to update a parameter μ, while other parameters are identical to those given by Equations 9.b to 9.d, since atomic adsorption only affects the inertia term dictated by a parameter μ. When atoms are locally adsorbed onto a graphene with adsorption site given as (x_m, y_m) as shown in Figure 1, the inertia term μ is given by

$$\begin{aligned}\mu &= \rho ab + \int_0^a \int_0^b \Delta m \delta(x - x_m)\delta(y - y_m)\psi^2(x, y)dxdy \\ &= \rho ab + \Delta m \psi^2(x_m, y_m)\end{aligned} \qquad (10.a)$$

where $\delta(x)$ is the Dirac delta function, and Δm is the total mass of adsorbed atoms. For the case in which atoms are uniformly adsorbed onto a graphene (i.e., mass adsorption occurs on the entire surface of a graphene resonator), the inertia term μ is written as

$$\begin{aligned}\mu &= \rho ab + \int_0^a \int_0^b \Delta m_0 \psi^2(x, y)dxdy \\ &= \rho ab + \Delta m_0 ab \equiv \rho ab + \Delta m\end{aligned} \qquad (10.b)$$

Here, Δm_0 is the mass of atoms adsorbed onto the unit area of a graphene, while Δm is the total mass of adsorbed atoms onto the entire surface of a graphene. Based on inertia terms for both mass-adsorbed graphene (as described in Equation 10) and a bare graphene (represented in Equation 9.a), it is straightforward to compute the resonant frequency shift, $\Delta\omega$, of a graphene resonator due to the mass adsorption; $\Delta\omega = \omega(m + \Delta m) - \omega(m)$, where $\omega(m)$ is the resonant frequency of a bare graphene resonator whose effective mass is given as m (with $m = \rho ab$), and $\omega(m + \Delta m)$ is the resonant frequency of a graphene resonator onto which the mass adsorption with the amount of Δm occurs. This frequency shift due to mass adsorption is typically negative since mass adsorption increases the overall mass of a resonator and, consequently, reduces the resonant frequency of a resonator with respect to that of a bare resonator.

3. RESULTS AND DISCUSSION

3.1 Resonance behavior of graphene

To verify the robustness of a continuum elastic model described in the section 'Theory and model', we have considered the vibration behavior of a graphene resonator with a size of 6 μm × 6 μm when a graphene is actuated by a force amplitude of p_0 = 0.001 aN (where 1 aN = 10^{-18} N) in order to induce the harmonic oscillation of a graphene resonator. Moreover, we have taken into account the case in which an in-plane tension N_0 is driven by pre-strain, such as $N_0 = E_s h\epsilon_0/(1 - \nu)$, where $E_S h$, and ν indicate the stretching modulus, thickness, and Poisson's ratio of a graphene resonator, respectively, and ϵ_0 is the pre-strain applied to a graphene resonator. With $\epsilon_0 = 4 \times 10^{-5}$, the resonant frequency of a graphene undergoing a harmonic oscillation is predicted as ω_0 = 20.63 MHz, which is consistent with the experimentally measured resonant frequency of 19.8 MHz (see Figure 2a and also ref. [11]). This indicates that a continuum elastic model (i.e., plate model) is suitable for understanding the dynamic behavior of a graphene resonator. It should be noted that a continuum elastic model overestimates the resonant frequency of a graphene when compared with that measured from experiments (for more details, see description as follows).

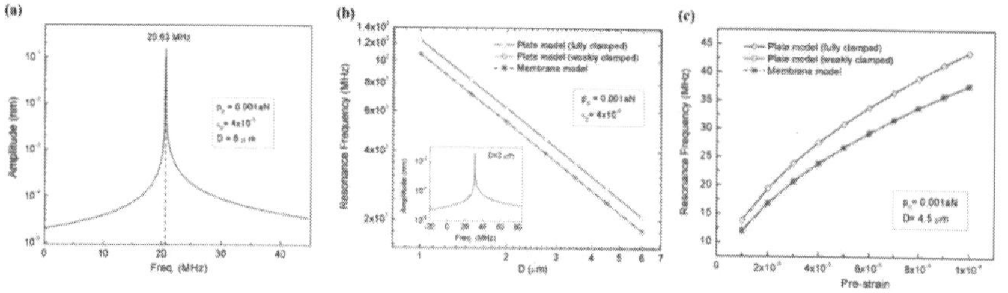

Figure 2 Resonance Behavior Of Graphene In Harmonic Oscillation. (**A**) Resonance curve of a squared graphene resonator with a size of D = 6 μm, where a graphene operates in harmonic oscillation driven by an actuation amplitude of $p_0 = 10^{-3}$ aN. (**B**) Resonant frequencies of graphene resonators operating in harmonic oscillation as a function of not only their size but also the boundary conditions. (**C**) Resonant frequencies of a graphene as a function of pre-strain applied to it.

As described above, the resonant frequency of a graphene resonator predicted from a continuum elastic model is overestimated in comparison with that measured from experiment. This may be attributed to the boundary condition that we used in our simulation; in our simulation, we have utilized the fully clamped boundary conditions such as $\psi(x\,y) = \nabla\psi(x\,y) = 0$ along all edges; this boundary condition is referred to as "fully clamped" boundary condition. In order to understand the effect of boundary condition on the resonant frequency of a graphene resonator, we have taken into account the deflection eigenmode in the form of $\psi(x\ y) = \sin(\pi x/a)\cdot\sin(\pi y/b)$, which satisfies the boundary condition as follows: $\psi(x\ y) = 0$ at all edges, but $\nabla\psi(x\ y) \neq 0$ along all edges. This boundary condition is referred to as "weakly clamped" boundary condition. As shown in Figure 2b, the resonant frequency of a graphene resonator that is weakly clamped is almost close to the theoretical predictions obtained from the membrane model [15]. Moreover, it is shown that the resonant frequency of a graphene resonator is critically dependent on the boundary condition; the weak clamping of a graphene resonator reduces its resonant frequency. In this work, we have considered a fully clamped graphene resonator, otherwise specified. The resonant frequencies of graphene resonators predicted from our continuum elastic model (i.e., plate model) are consistent with experimentally measured frequencies of graphene resonators [15]. In addition, it is shown that pre-strain increases the resonant frequency of a graphene (Figure 2c).

Now, we have taken into account the nonlinear oscillation of a squared graphene resonator with a size of $D = 250$ nm, to which pre-strain with the amount of $\epsilon_0 = 10^{-5}$ is applied. It is shown that when the amplitude of an actuation force is in the order of 0.01 fN, the graphene resonator undergoes harmonic oscillation. On the other hand, when a graphene resonator is actuated by the amplitude of an actuation force in the order of >0.05 fN, the graphene resonator experiences nonlinear vibration (Figure 3a). This is attributed to the geometric nonlinear effect due to fully clamped boundary condition. In order to quantitatively characterize the nonlinear vibration of a graphene resonator, we have introduced a dimensionless parameter $\theta = (\omega - \Omega_0)/\Omega_0$, where ω is the resonance of a nonlinearly oscillating graphene, and Ω_0 is the harmonic resonance of a graphene defined as $\Omega_0 = (\alpha/\mu)^{1/2}$. Here, it should be noted that the dimensionless parameter θ represents the degree of nonlinearity for the resonance of a graphene. Figure 3b shows the nonlinearity of graphene resonance (dictated by the dimensionless parameter θ) as a function of the amplitude of an actuation force. It is found that when a graphene resonator bears a pre-strain with the amount of 10^{-5}, the amplitude of an actuation force in the order of 0.5 fN results in θ = 0.15, which indicates that the resonance behavior at an amplitude of 0.5 fN is close to the harmonic oscillation. As the amplitude increases, the dimensionless parameter θ significantly increases, indicating that the nonlinearity of graphene resonance can be induced by a large amplitude of actuation force. In particular, when the graphene resonator is actuated with an actuation amplitude of 5 fN, the parameter θ becomes θ = 0.75, indicating that the resonance behavior is highly nonlinear. This indicates that the nonlinear vibration of a graphene resonator can be easily observed even when the actuation amplitude is applied in the order of 1 fN. The nonlinear vibration of a graphene resonator actuated with an actuation amplitude p_0 even in the order of 1 fN is attributed to the fact that the deflection amplitude of a graphene resonator is typically much larger than the thickness of a graphene resonator.

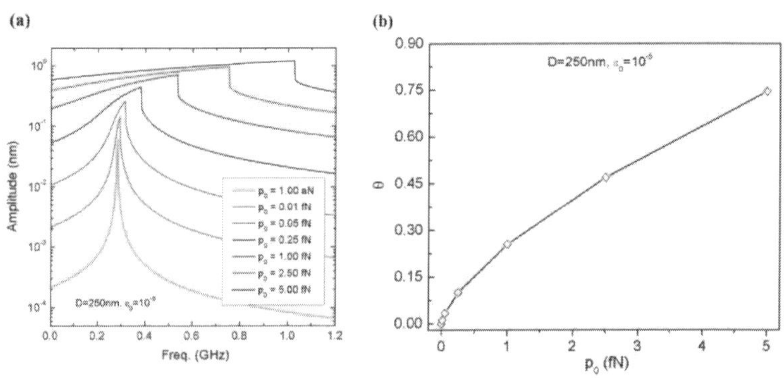

Figure 3 Resonance Behavior Of Graphene In Nonlinear Oscillation. (**A**) Resonance curves of a graphene resonator (with its size of D = 250 nm) as a function of actuation amplitude p_0. (**B**) Dimensionless parameter θ, which represents the degree of nonlinearity in the vibration of a graphene resonator, as a function of actuation amplitude p_0.

3.2 Resonant response of graphene resonators to mass adsorption

Even though graphene resonator has recently been extensively considered as a NEMS device for its applications in actuation, it has been barely employed as a mass sensor that enables highly sensitive atomic detection (e.g., measurement of atomic weight). The high detection sensitivity of a graphene resonator is attributed to the high-frequency dynamic range that is achieved due to high elastic stiffness and low mass density of a graphene resonator. In this study, we have scrutinized the resonant responses of graphene resonators, which undergo not only harmonic oscillation but also nonlinear vibration, to atomic adsorption onto the surface of a graphene resonator. Our study on the nonlinear response of a graphene resonator to mass adsorption is ascribed to our previous finding [9, 20, 21] that nonlinear oscillation is useful in improving the detection sensitivity of a nanoresonator.

We have considered a squared graphene resonator whose size is D = 250 nm without any applied pre-strain (i.e., $N_0 = 0$). Figure 4a depicts the resonant frequency shift of the graphene resonator due to atomic adsorption as a function of atomic mass as well as the amplitude of an actuation force. In this work, we have assumed that atomic mass was adsorbed onto the center of a graphene resonator. In general, as shown in Additional file 1: Figure S1, the frequency shift due to mass adsorption for a graphene is dependent on the location at which atomic mass was adsorbed. Moreover, it is presumed that the stiffness of the adsorbed molecule is ignored since the elastic modulus of a graphene is in the order of 1 TPa [6], which is much higher than that of adsorbed molecules such as proteins whose elastic modulus is in the order of 10 GPa [42, 43]. It is shown that the resonant frequency shift due to mass adsorption is linearly proportional to the adsorbed mass when a graphene resonator is actuated by a small amplitude of actuation force (e.g., p_0 = 1 aN). On the other hand, when a graphene resonator is excited by a large amplitude of actuation force (e.g., p_0 = 5 fN), the frequency shift due to mass adsorption is no longer proportional to the adsorbed mass. It is also found that as the amplitude of actuation force increases, the resonant frequency shift due to mass adsorption significantly increases, which indicates that nonlinear oscillation increases the detection sensitivity of a graphene resonator as anticipated. In order to gain a deep insight into the effect of nonlinear oscillation on the detection sensitivity of a graphene resonator, we have taken into account the resonant response of a graphene resonator to atomic adsorption with the amount of Δm = 10 ag. As shown in Figure 4b, the frequency shift of a graphene resonator to atomic adsorption with the amount of Δm = 10 ag is critically dependent on the amplitude of actuation force. In addition, we have shown the aforementioned dimensionless parameter θ as a function of the amplitude of actuation force (Figure 4b). It is shown that when the actuation amplitude is <0.5 fN, which corresponds to the harmonic oscillation ($\theta < 0.2$), the frequency shift of a graphene resonator with a size of D = 250 nm (without any pre-strain) due to atomic adsorption with the amount of 10 ag is in the order of 0.1 GHz. On the other hand, when an actuation amplitude is increased to 5 fN, which corresponds to highly nonlinear oscillation (i.e., θ = 1.6), the frequency shift due to mass adsorption (with the amount of 10 ag) increases by about sixfold (i.e., $\Delta\omega$ = 0.6 GHz). This clearly elucidates that the nonlinear vibration is useful in increasing the frequency shift of a graphene resonator due to mass adsorption, which highlights the nonlinear vibration that improves the detection sensitivity of a graphene resonator.

We have also studied the effect of graphene size (D) on the frequency shift of a graphene resonator due to mass adsorption, $\Delta\omega$, with respect to the actuation amplitude (p_0). It is found that for a graphene resonator whose size is $D \geq 150$ nm, the increase of actuation amplitude enhances the frequency shift of a graphene resonator due to mass adsorption, which is consistent with our conjecture that nonlinear vibration improves the detection sensitivity of a graphene resonator. On the other hand, for a graphene resonator whose size is $D \leq 100$ nm, an increase in the actuation amplitude does not significantly amplify the frequency shift of a graphene resonator due to mass adsorption in comparison to that of a large-scale graphene resonator (i.e., $D > 200$ nm). This indicates that even though large actuation amplitude induces the nonlinear oscillation of a small-scale graphene resonator (e.g., $D < 100$ nm), the nonlinear vibration does not remarkably increase the frequency shift due to mass adsorption. This result suggests that the length scale of a graphene resonator plays a central role in not only the dynamic frequency range of a graphene resonator but also in the sensing performance of graphene-based nonlinear oscillators.

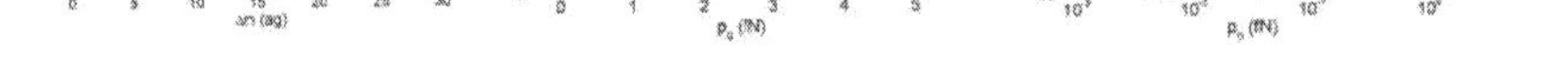

Figure 4. Graphene Resonators Response To Mass Adsorption. (**A**) Frequency shifts of a graphene resonator due to mass adsorption as a function of actuation amplitude p_0 and the amount of adsorbed mass Δm. (**B**) Frequency shifts of a graphene resonator due to atomic adsorption with the mass of Δm = 10 ag are found as a function of actuation amplitude p_0, while the dimensionless parameter θ of a graphene resonator with mass adsorption of Δm = 10 ag is shown. (**C**) Frequency shifts of graphene resonators due to atomic adsorption with mass of Δm = 20 ag with respect to the size of graphene resonators as well as the actuation amplitude p_0.

3.3 Effect of pre-strain applied to graphene resonators on their resonance behaviors and sensing performances

As described in previous studies [9, 20, 21], a mechanical tension (due to pre-strain or pre-stress) applied to a resonator leads to the increase of the dynamic frequency range of a resonator as well as its sensing performance. In this study, we have investigated how a pre-strain applied to a graphene resonator improves not only the dynamic behavior of a graphene resonator but also the detection sensitivity of a graphene resonator that operates in both harmonic and nonlinear oscillations.

We have studied the frequency change of a graphene resonator due to an in-plane tension with respect to the actuation amplitude (Figure 5a). Here, the frequency change is defined as the difference between the resonant frequencies of a graphene resonator bearing an in-plane tension and a bare graphene resonator, respectively. For a graphene resonator operating in harmonic oscillation, an in-plane tension increases the resonant frequency of a graphene resonator, which attributes to the fact that an in-plane tension stiffens the system. When a graphene resonator is actuated by the actuation amplitude in the order of 1 fN (leading to the nonlinear vibration of a graphene), it is interestingly found that the application of an in-plane tension (<6 pN/nm) to a graphene resonator reduces the resonant frequency of a graphene, which indicates that the in-plane tension is not useful in increasing the dynamic frequency range of a graphene operating in nonlinear vibration. This is consistent with our previous studies [20, 21] reporting that the dynamic frequency range of a nanoresonator undergoing nonlinear oscillation is decreased by a mechanical tension. However, it is remarkably found that when an in-plane tension is >6 pN/nm, the application of such in-plane tension increases the resonant frequency of a graphene resonator actuated by the amplitude of 1 fN. This may be attributed to the conjecture that an in-plane tension of 6 pN/nm to a graphene resonator actuated by an amplitude of 1 fN may induce the transition from nonlinear vibration to harmonic oscillation.

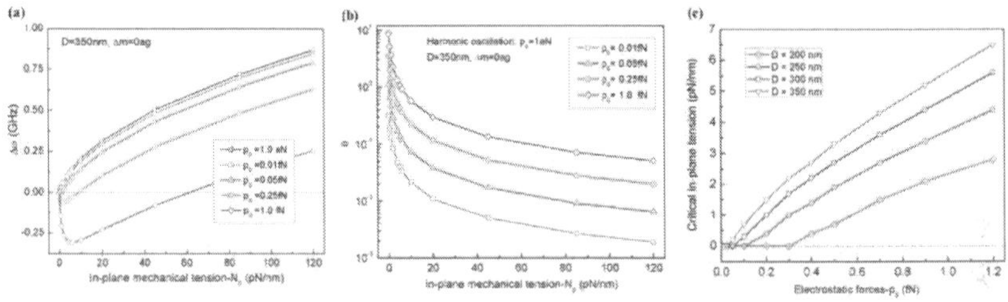

Figure 5. Effects Of Pre-Strain On Bare Graphene Resonator. (A) Frequency shifts of a bare graphene resonator (i.e. without any mass adsorption) due to in-plane tension as a function of an in-plane tension N_0 and actuation amplitude p_0. **(B)** Dimensionless parameter θ of graphene resonators as a function of actuation amplitude p_0 and in-plane tension N_0. **(C)** Critical in-plane tensions, which induce the transition from nonlinear vibration to harmonic oscillation for graphene resonators, are computationally obtained as a function of the size of a graphene resonator and actuation amplitude.

In order to validate our conjecture that an in-plane tension can play a role in the transition from nonlinear vibration to harmonic oscillation, as shown in Figure 5b, we have plotted the dimensionless parameter θ (representing the degree of nonlinearity) as a function of actuation amplitude and in-plane tension. When the actuation amplitude is in the order of 0.01 fN, the vibration behavior of a graphene becomes almost harmonic oscillation regardless of in-plane tension, as anticipated. For a graphene resonator actuated by an actuation amplitude of 1 fN, the dimensionless parameter for a bare graphene resonator (i.e. N_0 = 0) is in the order of 10, which indicates that the vibration behavior of a bare graphene is almost nonlinear oscillation. On the other hand, when an in-plane tension with the amount of 10 pN/nm is applied to a graphene resonator actuated by the amplitude of 1 fN, the dimensionless parameter is in the order of 0.5, which indicates that an in-plane tension reduces the nonlinearity of the resonance of a graphene. As in-plane tension increases, the nonlinearity is significantly reduced even up to the order of 10^{-1}, indicating that the resonance behavior is almost harmonic oscillation. Our result suggests that an in-plane tension plays a central role not only in increasing the dynamic frequency range of a graphene resonator but also in inducing the transition from nonlinear vibration to harmonic oscillation.

Figure 5c depicts the critical in-plane tension that is responsible for the transition from nonlinear resonance to harmonic oscillation. Here, the critical in-plane tension is defined as an in-plane tension at which nonlinear oscillation is transitioned to harmonic vibration. It is found that the role of in-plane tension in such transition is highly correlated with the size of a graphene resonator. For instance, for a graphene resonator whose size is D = 200 nm, the resonance behavior of a graphene actuated by an amplitude of ≤0.3 fN is almost close to harmonic oscillation. On the other hand, the vibrational behavior of a 350-nm graphene resonator driven by even an amplitude of 0.1 fN becomes nonlinear oscillation. This indicates that the size of a graphene resonator determines the actuation amplitude that is required to induce the nonlinear vibration of a graphene resonator. Moreover, it is shown that the smaller the graphene resonator is, the smaller is the amount of an in-plane tension that can induce the transition from nonlinear oscillation to harmonic resonance. This suggests that an in-plane tension-driven transition from nonlinear vibration to harmonic oscillation is determined by the size of a graphene resonator.

Now, we have studied the role of an in-plane tension on the detection sensitivity of a graphene resonator that experiences both nonlinear vibration and harmonic oscillation (Figure 6). For a graphene resonator operating in harmonic oscillation (e.g., a graphene actuated by an amplitude of 1 aN), an in-plane tension critically amplifies the frequency shift of a graphene resonator due to mass adsorption, which is consistent with our conjecture that the detection sensitivity of a graphene resonator is increased by an in-plane tension due to the in-plane tension-driven increase of the resonant frequency of a graphene. On the other hand, for a graphene resonator undergoing nonlinear vibration (e.g., actuated by an actuation amplitude of 1fN), an in-plane tension with the amount of <7 fN (corresponding to the critical in-plane tension that induces the transition from nonlinear vibration to harmonic oscillation of a graphene) decreases the amount of frequency shift due to mass adsorption, which suggests that an in-plane tension is ineffective in improving the detection sensitivity of a graphene resonator operating in nonlinear oscillation. However, when a graphene resonator is actuated by an amplitude of >7 fN, an in-plane tension increases the amount of frequency shift for a graphene resonator due to mass adsorption, which is attributed to the fact that a graphene resonator actuated by an amplitude of >7 fN obeys the harmonic oscillation. Moreover, we have also investigated the frequency shift of a graphene resonator, which operates in either harmonic oscillation or nonlinear vibration, due to mass adsorption (i.e., Δm = 20 ag) as a function of the size of graphene as well as in-plane tension. It is interestingly found that for graphene resonators operating in both nonlinear oscillation and harmonic vibration, an in-plane tension-induced improvement of detection sensitivity of a graphene resonator is significantly dependent on the size of a graphene such that an in-plane tension is useful in increasing the detection sensitivity of a graphene resonator whose size is $D \approx$ 100 nm, whereas an in-plane tension is ineffective in enhancing the sensing performance of a graphene with a size of $D >$ 300 nm in comparison with the detection sensitivity of a graphene resonator with a size of D = 100 nm. Our study sheds light on the important role of an in-plane tension on modulating not only the resonance behavior of a graphene resonator but also the detection sensitivity of a graphene resonator.

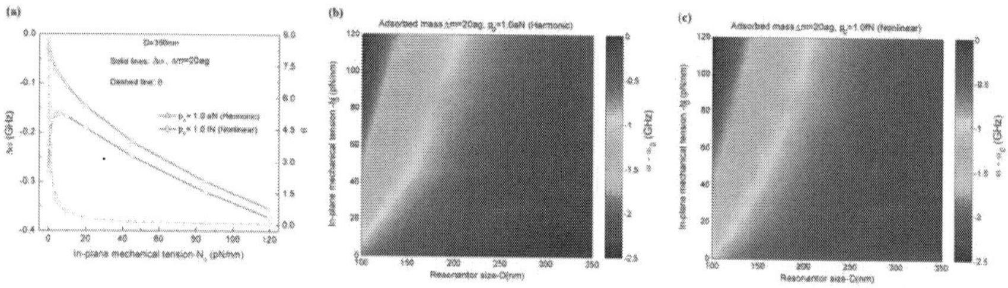

Figure 6. Effects Of Pre-Strain On Graphene Resonator In Either Harmonic Or Nonlinear Oscillation. (**A**) Frequency shifts of a graphene resonator, operating in either harmonic oscillation or nonlinear vibration, due to atomic adsorption with mass of Δm = 20 ag as a function of in-plane tension. (**B**) Frequency shifts of graphene resonators, which undergo harmonic oscillations, due to atomic adsorption (with mass of Δm = 20 ag) as a function of the size of graphene resonator and in-plane tension. (**C**) Frequency shifts of graphene resonators, operating in nonlinear vibrations, due to atomic adsorption (with mass of Δm = 20 ag) with respect to the size of graphene and in-plane tension.

4. CONCLUSIONS

In this work, we have studied the vibrational behaviors of graphene resonators as well as their sensing performance based on continuum elastic model such as plate model. It is shown that nonlinear vibration is useful in improving the detection sensitivity of a graphene resonator and that an in-plane tension is able to tune both the nonlinearity of the vibrating graphene resonators and their detection sensitivity. It should be noted that, in this work, our continuum model is only applicable to a monolayer graphene resonator. For modeling the multilayered graphene resonator, the interactions between graphene sheets have to be considered in the continuum modeling [27], which will be studied for our future work. Moreover, our continuum elastic model discards the finite size effect (i.e., edge effect) on the dynamic behavior (and also sensing performance) of a monolayer graphene resonator; here, edge (stress) effect arises from the imbalance between coordination numbers for edge atoms and bulk atoms, respectively [25, 44]. This edge stress effect on a monolayer graphene is conceptually identical to the surface stress effect on a nanowire resonator [9]. Such edge effect on the frequency behavior of a graphene resonator and its sensing performance will be studied for our future work.

ACKNOWLEDGMENTS

This work was supported by the National Research Foundation (NRF) of Korea under grant numbers NRF-2010-0026223 and 2012R1A1A2008616 (to KE). One of the authors (C-WK) gratefully acknowledges the support from NRF under grant numbers 2012R1A2A2A04047240 and 2012R1A1A2008870.

COMPETING INTERESTS

The authors declare that they have no competing interests.

AUTHORS' CONTRIBUTIONS

KE and C-WK designed the research. MDD performed the simulation. MDD, C-WK, and KE analyzed the data and wrote the manuscript. All authors read and approved the final manuscript.

REFERENCES

1. Geim AK, Novoselov KS: The rise of graphene. Nat Mater 2007, 6: 183. 10.1038/nmat1849

2. Castro Neto AH, Guinea F, Peres NMR, Novoselov KS, Geim AK: The electronic properties of graphene. Rev Mod Phys 2009, 81: 109. 10.1103/RevModPhys.81.109

3. Geim AK: Nobel lecture: random walk to graphene. Rev Mod Phys 2011, 83: 851. 10.1103/RevModPhys.83.851View Article

4. van den Brink J: Graphene: from strength to strength. Nat Nanotechnol 2007, 2: 199. 10.1038/nnano.2007.91

5. Wei X, Fragneaud B, Marianetti CA, Kysar JW: Nonlinear elastic behavior of graphene: ab initio calculation to continuum description. Phys. Rev. B 2009, 80: 205407.

6. Lee C, Wei X, Kysar JW, Hone J: Measurement of the elastic properties and intrinsic strength of monolayer graphene. Science 2008, 321: 385. 10.1126/science.1157996

7. Jiang J-W, Wang J-S, Li B: Young's modulus of graphene: a molecular dynamics study. Phys. Rev. B 2009, 80: 113405.

8. Cadelano E, Palla PL, Giordano S, Colombo L: Nonlinear elasticity of monolayer graphene. Phys Rev Lett 2009, 102: 235502.

9. Eom K, Park HS, Yoon DS, Kwon T: Nanomechanical resonators and their applications in biological/chemical detection: nanomechanics principles. Phys. Rep. 2011, 503: 115. 10.1016/j.physrep.2011.03.002

10. Poot M, van der Zant HSJ: Mechanical systems in the quantum regime. Phys. Rep. 2012, 511: 273. 10.1016/j.physrep.2011.12.004

11. Bunch JS, van der Zande AM, Verbridge SS, Frank IW, Tanenbaum DM, Parpia JM, Craighead HG, McEuen PL: Electromechanical resonators from graphene sheets. Science 2007, 315: 490. 10.1126/science.1136836

12. Garcia-Sanchez D, van der Zande AM, Paulo AS, Lassagne B, McEuen PL, Bachtold A: Imaging mechanical vibrations in suspended graphene sheets. Nano Lett 2008, 8: 1399. 10.1021/nl080201h

13. Chen C, Rosenblatt S, Bolotin KI, Kalb W, Kim P, Kymissis I, Stormer HL, Heinz TF, Hone J: Performance of monolayer graphene nanomechanical resonators with electrical readout. Nat Nanotechnol 2009, 4: 861. 10.1038/nnano.2009.267

14. Barton RA, Ilic B, van der Zande AM, Whitney WS, McEuen PL, Parpia JM, Craighead HG: High, size-dependent quality factor in an array of graphene mechanical resonators. Nano Lett 2011, 11: 1232. 10.1021/nl1042227

15. van der Zande AM, Barton RA, Alden JS, Ruiz-Vargas CS, Whitney WS, Pham PHQ, Park J, Parpia JM, Craighead HG, McEuen PL: Large-scale arrays of single-layer graphene resonators. Nano Lett 2010, 10: 4869. 10.1021/nl102713c

16. Kim SY, Park HS: On the utility of vacancies and tensile strain-induced quality factor enhancement for mass sensing using

graphene monolayer. Nanotechnology 2010, 21: 105710. 10.1088/0957-4484/21/10/105710

17. Eichler A, Moser J, Chaste J, Zdrojek M, Wilson-Rae I, Bachtold A: Nonlinear damping in mechanical resonators made from carbon nanotubes and graphene. Nat Nanotechnol 2011, 6: 339. 10.1038/nnano.2011.71

18. Hod O, Scuseria GE: Electromechanical properties of suspended graphene nanoribbons. Nano Lett 2009, 9: 2619. 10.1021/nl900913c

19. Atalaya J, Isacsson A, Kinaret JM: Continuum elastic modeling of graphene resonators. Nano Lett 2008, 8: 4196. 10.1021/nl801733d

20. Dai MD, Eom K, Kim C-W: Nanomechanical mass detection using nonlinear oscillations. Appl Phys Lett 2009, 95: 203104. 10.1063/1.3265731

21. Dai MD, Eom K, Kim C-W: Nanomechanical mass detection using nonlinear oscillation. Appl Phys Lett 2009, 95: 203104. A published erratum appears in Appl Phys Lett 2012, 100:179901 A published erratum appears in Appl Phys Lett 2012, 100:179901 10.1063/1.3265731

22. Kim SY, Park HS: The importance of edge effects on the intrinsic loss mechanisms of graphene nanoresonators. Nano Lett 2009, 9: 969. 10.1021/nl802853e

23. Kim SY, Park HS: Multilayer friction and attachment effects on energy dissipation in graphene nanoresonators. Appl Phys Lett 2009, 94: 101918. 10.1063/1.3099932

24. Qi Z, Park HS: Intrinsic energy dissipation in CVD-grown graphene nanoresonators. Nanoscale 2012, 4: 3460. 10.1039/c2nr30493g

25. Shenoy VB, Reddy CD, Ramasubramaniam A, Zhang YW: Edge-stress-induced warping of graphene sheets and nanoribbons. Phys Rev Lett 2008, 101: 245501.

26. Kitipornchai S, He XQ, Liew KM: Continuum model for the vibration of multilayered graphene sheets. Phys. Rev. B. 2005, 72: 075443.

27. Wilber JP, Clemons CB, Young GW, Buldum A, Quinn DD: Continuum and atomistic modeling of interacting graphene layers. Phys. Rev. B. 2007, 75: 045418.

28. Atalaya J, Kinaret JM, Isacsson A: Nanomechanical mass measurement using nonlinear response of graphene membrane. EPL 2010, 91: 48001. 10.1209/0295-5075/91/48001

29. Sadeghi M, Naghdabadi R: Nonlinear vibrational analysis of single-layer graphene sheets. Nanotechnology 2010, 21: 105705. 10.1088/0957-4484/21/10/105705

30. Timoshenko S: Theory of Plates and Shells. McGraw-Hill, New York; 1940.

31. Yang JZ, We : Generalized Cauchy-Born rules for elastic deformation of sheets, plates, and rods: derivation of continuum models from atomistic models. Phys. Rev. B. 2006, 74: 184110.

32. Erickson JL: Mathematics and Mechanics of Solids. 2008, 13: 199. 10.1177/1081286507086898

33. Arroyo M, Belytschko T: Finite crystal elasticity of carbon nanotubes based on the exponential Cauchy-Born rule. Phys. Rev. B 2004, 69: 115415.

34. DiVincenzo DP: Dispersive corrections to continuum elastic theory in cubic crystals. Phys. Rev. B 1986, 34: 5450. 10.1103/PhysRevB.34.5450

35. Mura T, Koya T: Variational Methods in Mechanics. Oxford University Press, Oxford; 1992.

36. Oden JT, Reddy JN: An Introduction to the Mathematical Theory of Finite Elements. Wiley, New York; 1976.

37. Bathe K-J: Finite Element Procedures. Prentice Hall, Eaglewood Cliffs; 1996.

38. Zienkiewicz OC, Taylor RL: The Finite Element Method: Basic Formulation and Linear Problems. McGraw-Hill, London; 1989.

39. Dykman MI, Soskin SM, Krivoglaz MA: Spectral distribution of nonlinear oscillator performing Brownian motion in a double-well potential. Physica A. Statistical and Theoretical Physics 1985, 133: 53. 10.1016/0378-4371(85)90056-1

40. Dykman MI, Mannella R, McClintock PVE, Moss F, Soskin SM: Spectral density of fluctuations of double-well Duffing oscillator driven by white noise. Phys Rev A 1988, 37: 1303. 10.1103/PhysRevA.37.1303

41. Greenberg MD: Foundation of Applied Mathematics. Prentice Hall, Eaglewood Cliffs; 1978.

42. Yoon G, Park H-J, Na S, Eom K: Mesoscopic model for mechanical characterization of biological protein materials. J Comput Chem 2009, 30: 873. 10.1002/jcc.21107

43. Yoon G, Kwak J, Kim JI, Na S, Eom K: Mechanical characterization of amyloid fibrils using coarse-grained normal mode analysis. Adv Funct Mater 2011, 21: 3454. 10.1002/adfm.201002493

44. Reddy CD, Ramasubramaniam A, Shenoy VB, Zhang Y-W: Edge elastic properties of defect-free single-layer graphene sheets. Appl Phys Lett 2009, 94: 101904. 10.1063/1.3094878

CHAPTER 5

Influence of the Delay and Dispersion in Mechanics

Evelina V. Prozorova

Mathematics & Mechanics Faculty, St. Peterburg State University, St. Peterburg, Russia

ABSTRACT

The aim of this work is to clarify the new mathematical model describing the mechanics of continuous media and rarefied gas. The present study is associated with the formulation of conservation laws as conditions of equilibrium of angular momentums, while usually formulated in terms of balance of force. The equations for gas are found from the modified Boltzmann equation and the phenomenological theory. For a rigid body, the equations used the phenomenological theory, but changed their interpretation. We elucidate the contribution of cross-effects in the conservation laws of continuum mechanics, including the self-diffusion, thermal diffusion, etc., which indicated S. Wallander. The paradox of Hilbert in the solution of the Boltzmann equation by the Chapman-Enskog method was resolved. Refined model of the boundary conditions for rarefied gas flows and transient flow were near the moving surfaces. We establish conditions for the existence of the A. N. Kolmogorov inertial range on the basis of the proposed theory. Based on the theory, derivation of the Prandtl formula for boundary layer was received. Delay in mechanics plays an important role on commensurability of relaxation times and lateness. New accounting delay option is proposed to consider the difference between the time derivative as a limit and end values of the mean free path in a rarefied gas. The role of individual time delay for each particle velocity and the average time is debated. The Boltzmann equation is written with an additional term. This situation is typical for discrete medium. The transition from discrete to continuous environment is a key issue mechanics. Summary records of all effects lead to a cumbersome system of equations and therefore require the selection of main effects in a particular situation. The role of the time has similar problems in quantum mechanics. Some examples are suggested.

KEYWORDS

Angular Momentum Conservation Laws, Unbalanced Stress Tensor, The Boltzmann Equation, Chapman-Enskog Method

1. INTRODUCTION

The proposed work involves new mathematical modeling of processes in the mechanics of fluids, gases and solids in classical and quantum cases. I agree with Stanislaw Jerzy Lec: I think that the amount of zeros is a formidable figure. It seems that continuum mechanics evolving for centuries by many luminaries of science has acquired a finished look in the formulation of its basic laws, but it is not. To date, we have complete mathematical description of certain physical phenomena insufficiently. Some of the issues will be discussed in this paper. The main question is the ratio of discrete and continuous descriptions. One of these effects will be discussed in this work. Previous studies have examined the effects of the angular momentum influence and the associated with its rotation of the elementary volume on writing of the conservation laws of continuum mechanics and kinetic theory. [1] repeated the analysis of previous work. Since gas often significantly alters the properties on small distance, the equations of the state are not sufficiently accurate to describe the structure of flow and solids. Dissipative effects occur at the borders, including on the contact surfaces through the formation and growth of the boundary layers and form turbulence. A large number of experiments tell us about the bad turbulence theories, bad equations of state of water and other liquids, poor boundary conditions in the case of transient flow and poor models of formation of various structures in the solid if gradients and large cracks tell us about the importance of gradients of physical quantities (density, momentum, energy, angular momentum). Our classical models work well when the gradients of physical values are small. We modified the Liuville equation and Bogolubov equation. L. D. Landau, C de Groot, P. Mazur, I. Deyrmati, I. Prigogine, L. I. Sedov, A. A. Ishlinskii, D. D. Ivlev et al. point at the role of angular momentum. It is the most developed area in the momentum theory of elasticity (R. D. Mindlin, V. A. Palmov, A. G. Gorshkov, E. I. Starovoitov, A. V. Yarovoy, V. Levin, S. E. Kanaun, E. L. Aero, etc.). Any movement of an elementary volume of liquid at the moment can be considered as a result of the following motion: quasisolid motion that is translated with selecting pole, rotating motion around this pole and deformation motion. This theorem was proved by H. Gelmholtz. L. Prandtl formulated conception of hardplastic body as the theory of ideal plasticity. Usually, we do not take into account twist velocity. The angular momentum is responsible for the twist velocity. Thus, in the classical Newton mechanics we have four conservation laws: masses, liner momentum, energy and angular momentum. In continuous mechanics, we use only three first laws. However, the law of conservation of angular momentum requires a particular frame of reference and a particular radius vector from the origin to the elementary volume. In the classical approach, the law of conservation of angular momentum does not fit. For these formulations, actually, they postulated symmetry of the material frame of reference and, as a consequence, the symmetry of the stress tensor and the violation of the "continuity" of the environment. Mathematically, this fact follows from the choice of the conditions of equilibrium of forces as a condition of equilibrium. Selection of the conditions of equilibrium of momentum of forces leads to new formulations of equations. The kinetic theory does not save the situation. The law of angular momentum is not implemented in the Boltzmann equation. Classical theory, the second viscosity, is predicted. But usually, we suppose that it is in need to take into consideration for molecules with inner of degree of freedom or for dense gas. The modified kinetic theory gives the second viscosity for molecules without structure. As a result, asymmetric stress tensor is obtained and sets the method of accounting of the asymmetric tensor. It is found that the stress tensor is asymmetric for particles without structure as for kinetic as for phrnomenological theories. It was suggested the kinetic theory with the angular momentum is as an additional variable, using a more accurate asymptotic approximation for macro parameters with a resolution of Hilbert's paradox. We believed that elementary volume can be rotated around the axis of inertia or to be involved in the rotational movement. In both cases, the density of the flow is across the border changes as $\frac{\mathrm{d}(\rho u)}{\mathrm{d}r}\cdot(r'-r)+\cdots$ by the rotation of the elementary volume. The contribution of other components is small, taking into account the smallness of the volume and the absence of rotation on axis.

Additional issues arise when writing communications stress tensors with tensors of strain rate. Thus, we propose valuing the order of the dependence of the classical relations in this paper. The order equations were increased, requiring additional boundary conditions. If we are dealing with a continuous medium, the boundary condition at the outer edge of the boundary layer can be defined by the value of the vortex or the value of the vertical velocity on the surface you want to set the friction. The turbulent layer usually gives friction. A rarefied gas flow must be set apart beside the usual boundary conditions. In general, the formulation of the boundary conditions requires a further study. Adhesion conditions hold true on a large scale for the flow of Newtonian type. In addition to the angular momentum, the impact of large gradients in the collision integral is essential. Field description (continuum mechanics) implies a transition from a discrete to a continuous medium with aspiration of the distances between atoms (molecules) to zero. In rarefied gas, unusual situation arises when we use the ratio of the increment limit function to the increment of the argument to describe derivatives. It turns out that for recording the time derivative we can consider only the high-speed components as slow collisions do not have time to occur at finite mean free path (rarefied gas). One model was suggested by B. V. Aleekseev [2] . Therefore, there is a need to add the second derivative of the Taylor series for inclusion in the work of the other more slowly components or use the average value of the derivative. For a finite-length path of the molecules when the Knudsen number of order unity, in the collision integral must also be taken into account "delay." Additional term can be computed. Given the orders of magnitude, the additional integral can be used locally equilibrium distribution function. This means that you can find the new core of the Navier-Stokes equations. In all cases, the classical theory is considered a limiting case when the volume-to-point, but the material point (particle)-simple physical models in mechanics—the ideal body size, and rotation may be neglected. We can also assume infinitely small body size compared with other dimensions or distances within the assumptions of the problem considered. As it was said derivation of conservation laws in classical mechanics is based on the integral conservation laws, shrinking to a point. In the second transition to the classical theory of conservation laws is accomplished by using the density sum of delta functions on the difference $(x_i - x)$, similar to line up for the conservation laws of momentum and energy. Thus, the transition from a discrete medium and back is carried out without taking into account the spatial distribution of physical quantities, static points. A similar situation is observed in quantum mechanics. Some of the questions for the probability density function associated with the transition from quantum to classical mechanics are discussed. Perhaps the best option theory of continuous medium is discrete approach in the integral form for elementary volume, which requires further study. Some examples are suggested. The part of examples one can find in [3] -[5] .

2. SOME SINGULARITIES OF THE BOLTZMANN EQUATION

We try to receive the modified Navie-Stokes equations from kinetic theory with angular momentum. Common essential in the Bogolubov theory assumptions are: infinite volume is considered; potential forces of the interaction of two particles rapidly abate with increasing of distance between two particles; the appearance of a finite volume's effect is investigated when the influence of the boundary is unessential, gas is rarefied but has a finite average density. The description of movement of the particles is based on neglect of the variation angular momentum for particles in an elementary volume (in a small neighborhood of the considered one). This is correct for relaxation process for which theory was developed. Essential that the fourth term in formula (1) for nonhomogeneous gas and for nonequilibrium gas does not vanish, when $|r_j| \to \infty$ [6] -[10]

$$\frac{1}{V^s}\frac{\partial F_s}{\partial t} = -\frac{1}{V^s}\sum_{i=1}^{s}\frac{p_i}{m}\frac{\partial F_s}{\partial r_i} + \frac{1}{2V^s}\sum_{\substack{i,j=1\\(j\neq i)}}^{s}\frac{\partial\Phi(|r_i - r_j|)}{\partial r_i}\frac{\partial F_s}{\partial p_i}$$
$$+\frac{1}{V^{s+1}}\sum_{i=1}^{s}\int\sum_{j=s+1}^{N}\frac{\partial\Phi(|r_i - r_j|)}{\partial r_i}\frac{\partial F_{s+1}(t, r_1, \cdots, r_s, r_j, p_1, \cdots, p_s, p_j)}{\partial p_i}\mathrm{d}r_j\mathrm{d}p_j$$
$$-\frac{1}{V^{s+1}}\sum_{j=s+1}^{N}\int\frac{p_j}{m}\frac{\partial F_{s+1}(t, r_1, \cdots, r_s, r_j, p_1, \cdots, p_s, p_j)}{\partial r_j}\mathrm{d}r_j\mathrm{d}p_j$$
$$+\frac{1}{V^{s+1}}\sum_{i=1}^{s}\int\sum_{j=s+1}^{N}\frac{\partial\Phi(|r_i - r_j|)}{\partial r_j}\frac{\partial F_{s+1}(t, r_1, \cdots, r_s, r_j, p_1, \cdots, p_s, p_j)}{\partial p_j}\mathrm{d}r_j\mathrm{d}p_j$$
$$+\frac{1}{2V^{s+2}}\sum_{\substack{i,j=s+1\\(j\neq i)}}^{s}\int\frac{\partial\Phi(|r_i - r_j|)}{\partial r_i}\frac{\partial F_{s+2}(t, r_1, \cdots, r_s, r_i, r_j, p_1, \cdots, p_s, p_i, p_j)}{\partial p_i}\mathrm{d}r_i\mathrm{d}r_j\mathrm{d}p_i\mathrm{d}p_j$$

.

(1)

Usually we suggest that for $|r_j| \to \infty$ $\frac{\partial}{\partial r_j}F_n \to 0$, $\frac{\partial}{\partial p_j}F_n \to 0$ $|p_j| \to \infty$.

There are: t—time, r—coordinate, p—momentum, r—density, Φ—potential of molecules interaction. The last three terms are equal to zero at indicated conditions. For nonequilibrium case spatial derivatives do not equal to null. It does not equal to null in case, as example for interaction blowing gas from the surface with nonequilibrium gas of flow near the body. For the enumerated cases particles are not pairwise permutable and we does not have sum. Besides the flows of molecules have to include as the boundary conditions.

$$\int \sum_{j=s+1}^{N} \frac{\partial \Phi_{ij}}{\partial r_i} \frac{\partial F_{s+1}\left(t, r_1, \cdots, r_s, r_j, p_1, \cdots, p_s, p_j\right)}{\partial p_i} \mathrm{d}r_j \mathrm{d}p_j \approx N \int \frac{\partial \Phi_{is+1}}{\partial r_i} \frac{\partial F_{s+1}}{\partial p_i} \mathrm{d}r_{s+1} \mathrm{d}p_{s+1}$$

Essentially, we cannot receive equation for S-particle distribution function under great gradients of physical values by using ordinary approximations. Hence it is not possible to construct the Boltzman equation from the Hamilton formalism but we can use only the laws of conservation for mass, momentum and energy and angular momentum. We now construct of equation for N-particle distribution function for nonequilibrium case taking into account the great gradient of the physical values by using the ordinary notations, i.e. $\vec{\mathbf{r}}$ —radius-vector; $\vec{\mathbf{x}}$— coordinate of the point and according to definition function f_N in element of physical volume $\mathrm{d}\vec{\mathbf{x}}$ near the point $\vec{\mathbf{x}}$ in moment t probable number of molecules with velocity in element $\mathrm{d}\vec{\xi}$ near the $\vec{\xi}$ are equal to $f_N\left(t, \vec{\mathbf{x}}, \vec{\xi}\right) \mathrm{d}\vec{\xi} \mathrm{d}\vec{\mathbf{x}}$. Let us calculate the change of the angular momentum near the point $\vec{\mathbf{x}}$

$$\delta L = m(\mathbf{r} + \delta \mathbf{r}) \times f_N\left(t, \mathbf{r} + \delta \mathbf{r}, \xi\right) \cdot \xi - m\mathbf{r} \times f_N\left(t, \mathbf{r}, \xi\right) \cdot \xi ,$$

$$f_N\left(t, \mathbf{r} + \delta \mathbf{r}, \xi\right) = f_N\left(t, \mathbf{r}, \xi\right) + \delta \mathbf{r} \cdot \frac{\partial f_N}{\partial \mathbf{r}} .$$

It is necessary do because flow tube of an elementary volume must be order of free path and need to contain many molecules for ensuring of statistical property. Finally, we obtain the variation of angular momentum in any volume and this result is $\delta\mathbf{L} = m(\delta\mathbf{r}\times f_N + \mathbf{r}\times\delta f_N)\cdot\xi$.

Hence, it follows that

$$\delta\mathbf{L} = m\delta\mathbf{r}\times\left(f_N + \mathbf{r}\cdot\frac{\delta f_N}{\delta\mathbf{r}}\right)\cdot\xi$$

The convective operator is modified as

$$\frac{\mathrm{d}f_N}{\mathrm{d}t} = \frac{\partial f_N}{\partial t} + \xi_i\cdot\left[\frac{\partial f_N}{\partial x_i}\right] + \xi_i\cdot\frac{\partial}{\partial x_i}\left[x_j\frac{\partial f_N}{\partial x_j}\right] - \frac{X_i}{m}\frac{\partial f_N}{\partial \xi_i}.$$

Further we used this operator.

3. EQUATION

Presented in previous our papers the equations of motion, energy and angular momentum were obtained. In classical case we do not written the angular momentum equation. The modified equations are received by usually kinetic theory methods:

$$\frac{\partial \rho}{\partial t}+\frac{\partial \rho u_i}{\partial x_i}+\frac{\partial}{\partial x_i}\left(x_i\frac{\partial \rho u_i}{\partial x_i}\right)=0.$$

$$\frac{\partial \rho u_i}{\partial t}+\frac{\partial}{\partial x_i}\left(\rho u_i u_j+P_{ij}+x_i\frac{\partial P_{ij}}{\partial x_i}\right)-\frac{X_i}{m}\rho=0.$$

$$\frac{\partial}{\partial t}\rho\left(\frac{3}{2}RT+\frac{1}{2}u^2\right)+\frac{\partial}{\partial x_j}\left[\rho u_j\left(\frac{3}{2}RT+\frac{1}{2}u^2\right)+u_k P_{kj}+q_j\right]+\frac{\partial}{\partial x_i}x_i\frac{\partial}{\partial x_j}\left[\rho u_j\left(\frac{3}{2}RT+\frac{1}{2}u^2\right)+u_k P_{kj}+q_j\right]=0$$

$$\frac{\partial \vec{r}}{\partial x}\times\vec{p}_x+\frac{\partial \vec{r}}{\partial y}\times\vec{p}_y+\frac{\partial \vec{r}}{\partial z}\times\vec{p}_z+x_j\frac{\partial}{\partial x_j}\left(\vec{P}_j\right)=M_I$$

There are t—time, x, y, z—coordinate, ρ —density, Pi, j—stress tensor, u—velocity, q—thermal flow, R— gas constant. The last equation is to determine the degree of asymmetry of the stress tensor. The question arose when we are writing the law of conservation for density. Try to get it from phenomenological principles. The modified equation for density was received from kinetic theory in the form

$$\frac{\partial \rho}{\partial t}+\frac{\partial \rho u_i}{\partial x_i}+\frac{\partial}{\partial x_i}\left(x_i\frac{\partial \rho u_i}{\partial x_i}\right)=0.$$

Figure 1 shows that the linear velocity $\mathbf{u} = \boldsymbol{\omega} \times (\boldsymbol{r}' - \boldsymbol{r})$ is the velocity with respect to point M for kvazi-solid movement around axis r without forward velocity. The point M itself may be involved in the rotation. For an elementary volume $\mathbf{u} = \boldsymbol{\omega} \times (\boldsymbol{r}' - \boldsymbol{r})$ formula means a rotation around the axis of center of inertia but axis of moving of elementary volume can be lie outside it. So we have for twisting an elementary volume

$$\int_{(s)} \left(\nabla \rho \boldsymbol{u}(\boldsymbol{r}' - \boldsymbol{r}_c)\right)_n \mathrm{d}s = \int_{(s)} div(\nabla \rho \boldsymbol{u})(\boldsymbol{r}' - \boldsymbol{r}_c)\mathrm{dv}$$

The degree of asymmetric stress tensor we can received from momentum equation (in projections) ($\sigma = \tau$). The non-symmetric tensor τ is indicated.

$$x\left(\frac{\partial \sigma_{xz}}{\partial x} + \frac{\partial \sigma_{yz}}{\partial y} + \frac{\partial \sigma_{zz}}{\partial z}\right) - z\left(\frac{\partial \sigma_{xx}}{\partial x} + \frac{\partial \sigma_{yx}}{\partial y} + \frac{\partial \sigma_{zx}}{\partial z}\right) + \tau_{zx} - \tau_{xz} = 0$$

$$y\left(\frac{\partial \sigma_{xz}}{\partial x} + \frac{\partial \sigma_{yz}}{\partial y} + \frac{\partial \sigma_{zz}}{\partial z}\right) - z\left(\frac{\partial \sigma_{xy}}{\partial x} + \frac{\partial \sigma_{yy}}{\partial y} + \frac{\partial \sigma_{zy}}{\partial z}\right) + \tau_{zy} - \tau_{yz} = 0$$

$$x\left(\frac{\partial \sigma_{xy}}{\partial x} + \frac{\partial \sigma_{yy}}{\partial y} + \frac{\partial \sigma_{zy}}{\partial z}\right) - y\left(\frac{\partial \sigma_{xx}}{\partial x} + \frac{\partial \sigma_{yx}}{\partial y} + \frac{\partial \sigma_{zx}}{\partial z}\right) + \tau_{yx} - \tau_{xy} = 0$$

,

Another problem for the solving of the Boltzmann equation is the asymptotical methods. It is essential that selecting the local equilibrium distribution function f_0 as the basis in solution of the

Boltzmann equation by the Chapman-Enskog method exploits macroscopic parameters in f_0 from the Euler equations [10] . Macroscopic parameters are determined by the Chapman-Enskog. Distribution function leads to the Euler equations parameters and tensor P is symmetric. Formally in that way we have values (density, linear moment, energy) with mistake of the first order. This fact was noted by Hilbert without further use and correction.

$$\int \varphi(\xi) f^0 \mathrm{d}\xi = \int \varphi(\xi) f \mathrm{d}\psi = \beta,$$

there β-macroscopic parameters.

$$n(t,\vec{x}) = \int f(t,\vec{x},\vec{\xi})\mathrm{d}\vec{\xi}, \quad \boldsymbol{u}(t,\vec{x}) = \frac{1}{n}\int \vec{\xi} f(t,\vec{x},\vec{\xi})\mathrm{d}\vec{\xi},$$

$$P_{ij} = m\int c_j c_i f(t,\vec{x},\vec{\xi})\mathrm{d}\vec{\xi}, \quad q_i = \frac{m}{2}\int c^2 c_i f(t,\vec{x},\vec{\xi})\mathrm{d}\vec{\xi}.$$

So for the equilibrium in collision integral

$$f(t,x,\xi) \equiv f_0(t,x,\xi) = n\left(\frac{m}{2\pi kT}\right)^{3/2} \exp\left\{-\frac{m}{2kT}c^2\right\}, \quad \left(c^2 = c_1^2 + c_2^2 + c_3^2\right) = (\xi - \boldsymbol{u})^2$$

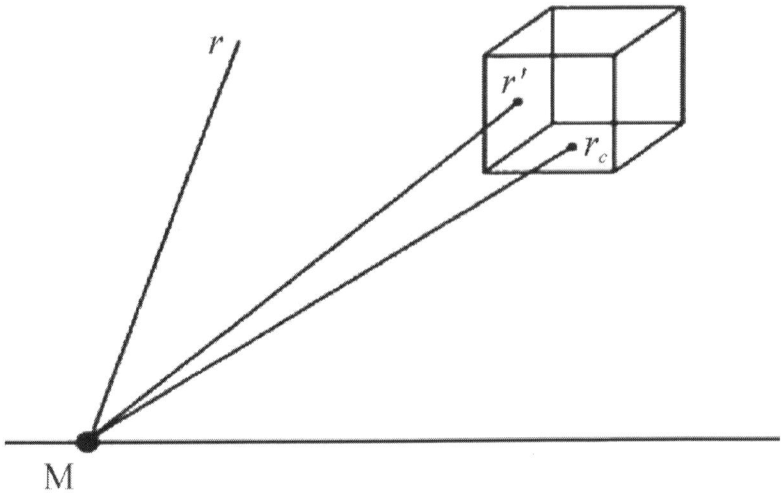

Figure 1. Change in density during the rotation of the elementary volume.

and for nonequilibrium distribution function

$$f = f_0\left[1+\frac{p_{ij}m}{2pkT}c_ic_j - \frac{q_im}{pkT}c_i\left(1-\frac{mc^2}{5kT}\right)\right]$$

we have the same macroparameters in f_0. The nonequilibrium distribution function is such that integral of it contains only the integral of equilibrium function f_0 and gives containing in f_0 macroscopic parameters. The remaining term gives null. Besides it is essentially for the value of viscosity. So it is necessary to do iteration all values (density, linear momentum, energy). It is possible for equation with angular momentum. In classic theory thus tensor P is entering as the value P in the Euler equations and as symmetric tensor in the Navier-Stokes equations. Consequently the theory value P for the Navie-Stokes equations is not equal to value P for the macroscopic Euler equations.

Also the angular momentum is very important characteristic for large gradients and the associated effects in the presence of charged particles. Now the angular momentum does not take consideration to calculate potential [11] . In the presence of interactions (that is to say, between different types of physical fields) quanta lost independence. Quantum theory of these fields does not sufficiently develop. Conserved angular momentum of a closed system and central forces, but in general, the law must be built. However, for the problem of decay of particles with mass defect after the collapse, we consider only the law of conservation of energy and momentum. In quantum mechanics separately makrozakon conservation of angular momentum is not issued. It is obtained by multiplying the vector equation for the amount of traffic on the vector r.

$$\frac{dJ_0}{dt} = -\nabla \cdot L, \text{ but it is not righ as } \frac{d(r \times v)}{dt} \neq r \times \frac{dv}{dt}.$$

Partially account the angular momentum occurs in solving the molecular dynamics method. Causality principle and the principle of the finite velocity of propagation of interactions require that the differential equations describing the fundamental fields belong to hyperbolic type. Despite the apparent first-order equations for the wave function and the equations of motion (for speed) together, we actually have a second-order equation and formally we can determine the velocity and momentum [12] [13] .

$$\nabla^2\Psi - \frac{1}{v^2}\frac{\partial^2\Psi}{\partial t^2} = 0,$$

$h\hat{L}$—Hamiltonian function or $ih\frac{d\Psi}{dt} = \hat{H}\Psi$; operator of density is $\rho(x',x) = \int \Psi^*(q,x')\Psi(q,x)dq$, the Wigner function of distribution [13] [14]

$$f_{lm,?m'}(\boldsymbol{r},\boldsymbol{p},t) = \frac{1}{(\pi h)^3}\int \exp\left(\frac{2i\boldsymbol{p}\cdot\boldsymbol{R}}{h}\right)\rho\left(\begin{array}{c|c} lm & lm' \\ r+R & r-R \end{array}\right)d^3\boldsymbol{R}$$

In continuum mechanics equations, were derived previously given angular momentum [1] . From this consideration, we need to consider angular momentum in quantum mechanics, i.e. taking as independent variables except the coordinate and momentum angular momentum. It will be a small correction, but for some options that have many interacting components, it can play a big role.

4. BOUNDARY LAYER

The Blasius problem was considered in [1] by numerical and analytical. New equations of uniform translational movement of cylinder are

$$\frac{\partial u}{\partial t}+u\frac{\partial u}{\partial x}+v\frac{\partial u}{\partial y}=U_e\frac{\partial U_e}{\partial x}+\tilde{v}\frac{\partial^2 u}{\partial y^2}+\tilde{v}\frac{\partial}{\partial y}\left(y\frac{\partial^2 u}{\partial y^2}\right),\ \tilde{v}=\frac{\mu}{\rho},$$

$$\frac{\partial u}{\partial x}+\frac{\partial v}{\partial y}+\frac{\partial}{\partial y}y\frac{\partial v}{\partial y}=0$$

Boundary conditions are

$$u=U_e(x),\quad v=0,\quad y=0,\ t=0;$$

$$u=0,\ v=0,\ \mu\frac{\partial u}{\partial y}=\tau_w,\ y=0;$$

$$u=U_e(x),\ y\to\infty,\ x>0;\quad u=U_e(0),\ x=0.$$

Here x, y are the coordinates, u, v are the velocities, $\frac{\mu}{\rho}$ is the viscosity, U_e is the velocity on external boundary. On figure two, a new result is presented for function of current for small indignation the type $U_e=U_\infty+\varepsilon\times\left(\frac{\partial u}{\partial y}-\frac{\partial v}{\partial x}\right)$, ε is small.

We have stretch team inside boundary layer. Result of numerical solution is on Figure 2.

Some results for infinite plate will be formulated here. The equation for this case is (the new term is selected)

$$\frac{d}{dy}\left(\mu\frac{du}{dy}\right)+\frac{d}{dy}\left(\mu y\frac{d^2u}{dy^2}\right)=0.$$

Boundary conditions are

$$u=0,\quad \mu\frac{du}{dy}=\tau_w,\quad y=0,$$
$$u=U_\infty,\quad y\to\infty.$$

Integrating gives

$$\mu\frac{du}{dy}+\mu y\frac{d^2u}{dy^2}=\text{Const}=\tau_w.$$

The Prandtl formula is received from this equation. From this equation we can receive the results of A. N. Kolmogorov.

We investigated the effect of the asymmetry of the distribution function for the equilibrium case, As since in directional velocity at the outer edge of a thin layer appears as a slow component, and high-energy. Therefore there is a walk along the surface of the particles to the slow component and the introduction of particles into the crystal for the fast component. It should be noted the possibility of a theoretical calculation of fatigue effects ideal surface in normal conditions. To do this, we need to calculate using the equilibrium distribution function of the value of the gas molecules with high energy for a given temperature, with a preliminary study, a similar representation.

The results we can see in [6] . One of results is on Figure 3.

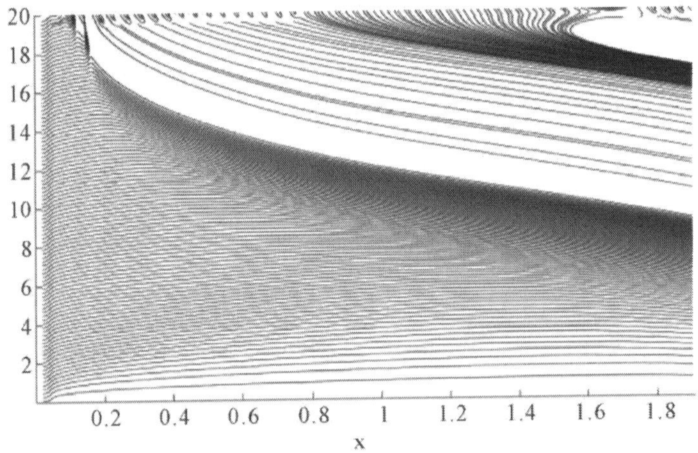

Figure 2. Stream function is inside the boundary layer at the vortex perturbation at the upper limit.

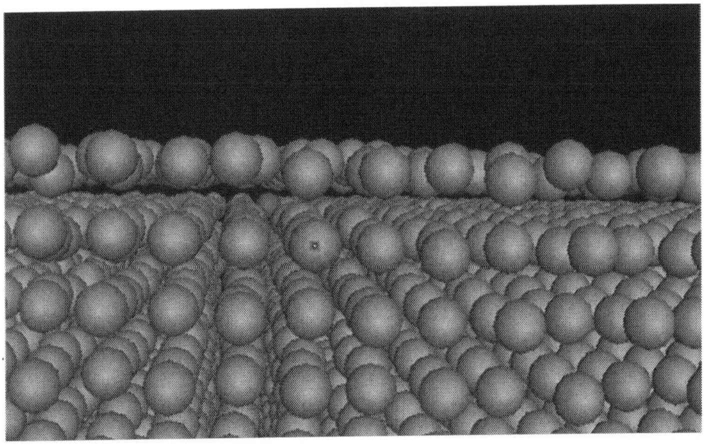

Figure 3. Reacting the directed gas stream with the equilibrium surface.

5. DELAY

In the kinetic theory, the role of delay in the considering should be to deal with the question of what values measure in the experiment: the instantaneous or averaged. If the experiment has to do with the average at the agreed time, in this case consider the delay is not necessary, except in cases of commensurability of relaxation times and the delay in otherwise, it is necessary to bear in mind the following: all interactions occur at a finite rate. Consider the action finite values of the propagation velocity of the interacting particles in the kinetic theory.

With that said, the Boltzmann equation can be written as

$$\frac{df}{dt}=\frac{\partial f}{\partial t}+\overline{\tau}\frac{\partial^2 f}{\partial^2 t}+c_i\cdot\frac{\partial f}{\partial r_i}+c_i\cdot\frac{\partial}{\partial r_i}r_{ij}\frac{\partial f}{\partial r_{ij}}-\frac{F}{m}\frac{\partial f}{\partial c_i}=I,\quad I=\Delta^- -\Delta^+$$

$$\Delta^-=\mathrm{d}t\,\mathrm{d}\vec{x}\,\mathrm{d}\vec{\xi}\,f\left(t,\vec{x},\vec{\xi}\right)\int\left[f_1\left(t,\vec{x},\vec{\xi}\right)+O\left(\Delta t\vec{\xi}\frac{\partial f_1}{\partial x}\right)\right]gb\mathrm{d}b\mathrm{d}\varepsilon\mathrm{d}\vec{\xi}_1$$

$$\Delta^+=\mathrm{d}t\,\mathrm{d}\vec{x}\,\mathrm{d}\vec{\xi}'\int\left[f\left(t,\vec{x},\vec{\xi}'\right)f\left(t,\vec{x},\vec{\xi}_1'\right)+O\left(\Delta t\vec{\xi}\frac{\partial f}{\partial x}\right)\right]g'b'\mathrm{d}b'\mathrm{d}\varepsilon'\mathrm{d}\vec{\xi}_1.$$

Similarly, calculated values with a stroke since the molecule must fly during the free path, the mean free path of the molecules and the travel time to and after the collision can be different (may be different values of the incident and impinging molecules (with index one).

However, structural gas, for example, at altitudes above 120 km the mean free time at three Mach numbers, i.e. lag time 10^{-8} c and more that can be comparable with the relaxation time. In fact, the expression can be simplified by considering the order of magnitude. Then

$$ff_1 - f'f_1' \leftrightarrow ff_1 - f'f_1' + \tau \frac{\partial f^0}{\partial t} f_1^0 + \tau_1 f^0 \frac{\partial f_1^0}{\partial t} + \lambda \frac{\partial f^0}{\partial x} f_1^0 + \lambda_1 \frac{\partial f_1}{\partial x} f^0 + \cdots - \tau' \frac{\partial f'^0}{\partial t} f_1'^0$$
$$- \tau_1' f^0 \frac{\partial f_1'^0}{\partial t} - \cdots - \lambda' \frac{\partial f'^0}{\partial x} f_1'^0 - \lambda_1' \frac{\partial f_1'^0}{\partial x} f' - \cdots$$

Then the integrals are computed and we obtain the corresponding kernel of the Navier-Stokes equations. For small and medium gradients mean free time one and one mean free path. But significant differences are in the interaction of gases with different properties.

6. CONCLUSIONS

The effect of the delay in a discrete medium under the transition to a continuous (field) description is investigated. A modification of the equations of a rarefied gas is done. Besides, we investigate the influence of angular momentum in the volume element. The influence of these effects in continuum mechanics and kinetic theory is showed. Some numerical results are demonstrated. Conditions of their application are studied and concrete situation are discussed.

Addition. All continuum mechanics have the same laws. So the theory elasticity needs to include the law of angular momentum. We suggest some examples.

Examples. If we shall consider protozoa task about beam and shall take into attention angular momentum, we shall receive the classical result. Consequently, we can choose the beam of variable cross-section. For beam on elastic foundation results distinguishes. In another case of transmission pressure at beam by intermediate designs of external force P (Figure 4) passes through hard lever of length l that can twist around hinge we show below. $H = \frac{P}{\cos\alpha}$. Here H- normal force in hard hinge.

At the end of elastic beam transfers force is $H = H\sin\alpha$. As $\mathrm{tg}\alpha = v(l)/a$ at l we have in classic case

$$Q = H\sin\alpha = N\,\mathrm{tg}\alpha$$

$$EIv'''(l) + N\left[v'(l) + \frac{v(l)}{a}\right] = 0.$$

So we have at end of length $v(0) = 0 \Rightarrow C_1 + C_4 = 0$,

$$v'(0) = 0 \Rightarrow C_2 + kC_3 = 0,$$

$$v'''(l) + k^2 v'(l) + k^2 v(l)/a = 0 \Rightarrow C_1 + C_2(l + a) + C_3 \sin v + C_4 \cos v = 0.$$

Now $EIv'''(l)+(Nv)'+\frac{P}{\cos\alpha}\frac{v(l)}{Q}=0$

If we write equations in classical case for stress values of elasticity theory, we have six unknown quantities and three equations. In our case, we add the equation for angular momentum but we have nine unknown quantities. As a result, we have old situation. For our case, we suppose that one possible solution we can receive is using the classical supposition. Consequently, we need to add equations for geometrical and physical properties. Now we consider four educational problems for long beam. The first one is the beam under longitudinal force. Suggested that

$$\sigma_x=\sigma_y=\tau_{xy}=0,\quad \sigma_z=\sigma_0=\frac{N}{\grave{o}}=\text{const};\quad \tau_{yz}=\tau_{zx}=0.$$

Here are the problems where three angular momentums are equal to zero. A new way to decide problem does not give new result. The second case is the clean torsion.

$$\sigma_z=ax+by,\quad \tau_{zy}=\tau_{zx}=0.\quad N=0,\ M_x=M_z=0.$$

Let us consider the problem where two angular momentums are equal to zero. Consequently, the laws of equilibrium forces are fulfilling. In this case, all three laws of force's balance must be equal to zero. For problem for transverse beam gives same results same for the second case. But for problem about twisting of the beam, we receive another solution $(\sigma_x = \sigma_y = \sigma_z = \tau_{xy} = 0)$. So two angular momentums at the end z = l are equal to zero.

We consider the Lame and Kirsha problems. In the first problem, loading on the sides is homogeneous. So momentum can be only in corner. For the same loading on all sides, angular momentum does not work. But for different loading, momentum works and added strength will be activity. The result is raising or going down. For the second problem besides the corner's effects, we have momentum near the axis z. So at cross of the circle and axis y, we have added forces that have the direction of the stress on side. So the new way to decide a problem does not give new results. The second problem is the clean torsion.

$$\sigma_z = ax + by,\ \tau_{zy} = \tau_{zx} = 0.\ N = 0,\ M_x = M_z = 0.$$

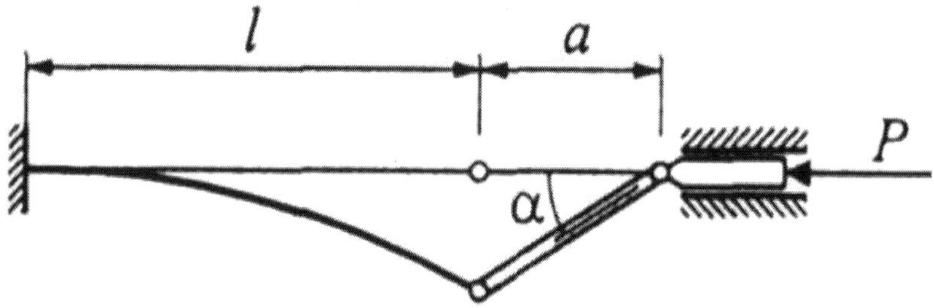

Figure 4. Transmission pressure at beam by intermediate designs of external force P.

Angular momentums are equal to zero. Consequently, the laws of forces are fulfilled. In this case all, three laws of force's balance must be equal to zero. For problem about transverse beam gives same results as for the second case. But for problem about twisting of the beam, we have another solution $\tau_{xy} = f(z)$ besides classic solution $(\sigma_x = \sigma_y = \sigma_z = \tau_{xy} = 0)$. We receive this result because two angular momentums are equal to zero. The above results are illustrated in Figure 5(a) and Figure 5(b) (for problems of Lama and Kirsch).

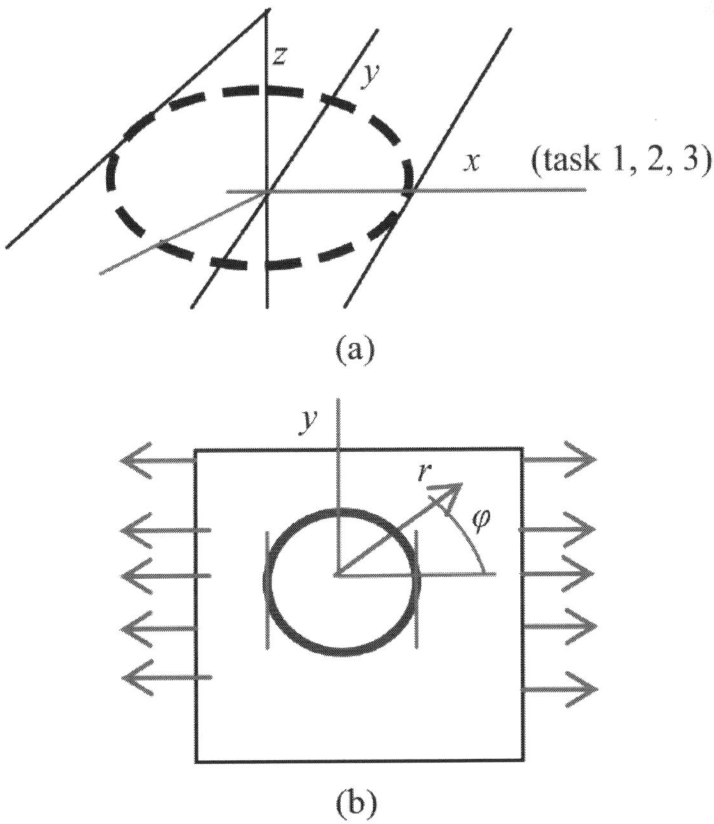

Figure 5. Illustrate for the problems of Lama and Kirsch.

REFERENCES

1. Prozorova, E.V. (2013) International Scientific IFNA-ANS Journal Problems of Nonlinear Analysis in Engineering Systems, 19, 45-57.

2. Alexeev (2004) Generalized Boltzmann Physical Kinetics. Elsevier, Amsterdam.

3. Prozorova, E.V. (2012) Electronic Journal Physico-Chemical Kinetics in Gas Dynamics, 13. http://www.chemphys.edu.ru/pdf/2012-10-30-001.pdf (in Russian)

4. Prozorova E.V. and Shadrin, A. (2013) Influence Dispersion in Gas and Solid for Moving Body. 5th European Conference for Aeronautics and Space Sciences, Munich, 1-5 July 2013/Holiday Inn Munich City Centre, Munich.

5. Prozorova, E.V. (2013) Influence of the Dispersion in Model of Continuous Mechanics. VVM Publishing Ltd., St.- Peterburg. (in Russian)

6. Bogolubov, N.N. (1946) Problems of Dynamic Theory in Statistical Physics. Gostexizdat, M-L. (in Russian)

7. Cercignani, C. (1969) Mathematical Methods in Kinetic Theory. Macmillan.

8. Ferziger, J.H. and Kaper, H.G. (1972) Mathematical Theory of Transport Processes in Gases. Amsterdam-London.

9. Hirschfelder, J.O., Curtiss, C.F. and Bird, R.B. (1954) The Molecular Theory of Gases and Liquids. New York.

10. Kogan, M.N. (1967) The Dynamics of the Rarefied Gases. Nauka, M. (in Russian)

11. Lazić, I. (2009) Atomic Scale Simulation of Oxide and Metal Film Growth. Chapter 3. General Performance Testing, Chapter 4. An Improved Molecular Dynamics Potential for Studying Aluminum Oxidation. Part1-Parameter Optimization for the Electrostatic Part of the Potential. College Voor Promoties.

12. Zubarev, D.N. (1971) Nonequilibrium Statistical Thermodynamics. Nauka, Moscow, 414 p.

13. Fermi, E. (1968) Quantum Mechanics. Wiley, New York.

14. Eugene, P. (1970) Wigner Symmetries and Reflections. Bloomigston-London.

CHAPTER 6

Modeling Fiber Composites during the Cure Process for Piezoelectric Actuation*

Darryl V. Murray, Oliver J. Myers

Department of Mechanical Engineering, Mississippi State University, Starkville, USA

ABSTRACT

Analytical, numerical, and experimental modeling methods are presented to predict deformation after the cure process of thin unsymmetric laminates for piezoelectric actuation. During fabrication, laminates deform to several post-cure room temperature shapes. Thin cross-ply laminates deform to a circular cylindrical post-cure shape while thicker laminates deform to a saddle shape. Post-cure shapes are dependent on ply orientation, thickness, and material properties. Because, CLT alone does not always predict the correct post-cure room temperature shape of the thin composite laminates, an extension of CLT with the Rayleigh-Ritz technique and potential energies are used to better predict these shapes. Finite element models are used to predict the post-cure room temperature shapes. Thin composite laminates are modeled coupling heat transfer and structural mechanics, which are necessary for modeling the cure process. Modeling the fabrication process captured important data such as residual stresses from the cure process, room temperature shapes, and bi-stability of the composite laminates. To validate these analytical and numerical results, experiments were conducted using macro-fiber composite (MFC) patches for morphing the laminates. The experimental piezoelectric morphing results relate well to analytical and numerical results.

KEYWORDS

Composite; Piezoelectric; Morphing; Modeling

1. INTRODUCTION

Smart material systems are vastly becoming an integral part in engineering applications. One of the phenomena used in smart material systems is piezoelectricity. Using unsymmetric bi-stable composites, piezoelectric effects can be implemented to achieve a snap through to the other stable shape of the composite [1]. Unsymmetric laminates are defined as laminates that are configured such that the geometric mid-plane is not a mirror image of the ply configurations above and below the mid-plane. This shape change caused by piezoelectric effects can be coupled with other domains and used as a sensor or actuator. To achieve bi-stability, a fabricated unsymmetric laminate is cured to a certain temperature then cooled to the operating or room temperature [1]. The curing process causes the laminate to deform due to the thermal strain gradient between the layers of the laminate. As an essential part of the piezoelectric system, the cure process and the resulting cured shapes are heavily investigated in the current paper.

After curing, a once flat laminate will deform into one of multiple shapes based on the laminate layup and the suggested cure temperature. Research and analysis are being done to predict the deformations and deflections of laminates. Dano and Hyer (2001) predicted the forces and moments that cause cure deformation using common theories based on displacements and strain fields [2]. Modeling three different lay-ups, the Rayleigh-Ritz method and the principal of work was used to determine these forces. Formulating equations for the first derivation of the strain energy, the reference or mid-plane strains and curvatures were approximated. These parameters were used to determine the in-plane and out-of-plane displacements using strain-displacement relationships. A system of equations was then used to approximate the work due to the thermal strains. The research continued on to applying moments to the laminate after the cure process to cause the laminate to snap through to the other stable shape. This research was done as an extension of Dano and Hyer (1996) previous work based on the same theories [3]. C. R. Bowen (2007) experimented with a cantilever beam of a [0/0/90/90] carbon fiber epoxy and an unsupported laminate of [0/90] for morphing testing and analysis [4]. The specimens were cured to a particular maximum temperature with constant pressure. Both specimens formed to one of the cylindrical shapes after the cure process. Piezoelectric actuation was then implemented using macro-fiber composite patches for electrical and mechanical coupling. Although the specific formulation of the cured deflections were not presented, the results were useful.

Hyer and Jilani (1998) presented a method for predicting the deformation of rectangular laminates cooled from the cure temperature to that of the room temperature [5]. The piezoceramic material named THUNDER was used for investigation. Due to the unsymmetric nature of the THUNDER system, deformation was observed as the system cooled from the cure temperature. After analyses were done, it was stated that the occurrence of one of the two stable shapes depends on the geometry of the laminate. This research consisted of also studying the issue of stability and what constitutes a laminate as being in one of the stable states.

The Classical lamination theory suggests that a laminate cooled from the cure temperature will achieve a saddle shape, the shape between the two stable shapes. This suggestion is not always correct. When cooled from the cure temperature, laminates can also attain one of the stable cylindrical shapes (concave up or down). M. Schlecht, K. Schulte, and Hyer (1995) extended the Classical lamination theory to approximate these shapes then compared these results to finite element analysis [6]. The results were used as parameters for the stress and strains during the snap through action to the other stable shape. It would seem that since geometry plays a role in the cured shape of the laminate, the ply orientation would also be a factor. D. N. Bettes, A. T. Salo, C. R. Bowen, and H. A. Kim (2010) investigates this along with other parameters effecting the cured shape [7].

Hyer also discusses the subject matter of stress analysis of fiber composites [8]. Hyer postulates many of the techniques used to determine stresses, strains, and displacements for a fiber composite including but not limited to these variables in reference to the cured shape. Numerous valuable techniques and assumptions can be found and have been used for this current research. Jones [9] also discusses the mechanics of composites used in this research and includes Hyer's work on post cure shape change deformation. The importance of the approximation of the cure shape and the cure deformation is the cause of the abundance of research done within this area of unsymmetric laminates.

M. Gigliotti and M. R. Wisnom studied the curvature due to the cure process of an AS4/8552 [0/90] laminate [10]. The cure process was broken down in to three stages and further investigation was done. It was hypothesized that inconsistent thermal properties between plies, chemical shrinkage of resin/epoxy, interaction of the tool or mold, and a non-uniform cure temperature are the four thermal stress-strains mechanisms that cause stress in the laminate during the cure process.

The snap through phenomenon for piezoelectric systems are of interest and will be investigated in further research; the present paper focuses on the curing process and approximating the stresses, strains, and curvatures of the unsymmetric laminate during this process. As Schlecht and Schulte studied the thermal effects of the cure process on room-temperature laminates, this paper establishes analytical and numerical modeling for stress, strain, and curvature prediction after the cure process [6]. Using carbon pre-preg sheets, the plies are cut to the specific ply orientation. Then using an aluminum mold and a releasing agent, the laminate is placed in the mold inside a convection oven. During the cure process, there are four mechanisms that cause curvatures and deformation in the laminate [11]. The development of stress in the cured laminate has a significant effect on the piezoelectric actuation and analysis. There can be more advances on the prediction of these variables for better understanding for the application of smart material systems. To achieve optimum snap through with piezoelectric materials, whether sensing or actuators types, it is best that the laminate possess curvatures and strains necessary for particular applications [12-14].

M. W. Hyer also used strain energy theories and virtual work formulation to predict the post cure displacements and curvatures of the laminate [3,5]. These methods were evaluated using various types of lay-ups and cure temperatures. M. W. Hyer formulated the necessary equations to predict the variables then compared these values to experimental values of the same laminate. The most common lay-up was that of [0/90] T and it was used due to the simplicity of the governing equations. Libo Ren and Azar Parvizi-Majidi investigated the CLT and how the predictions of the post cure shape were not always correct [15]. Studying cross-ply shells, Ren and Parvizi modeled the cure shape using the Rayleigh-Ritz energy method. This method allowed consideration of geometric nonlinearity, a problem with the general CLT. M. Gigliotti and M. R. Wisnom along with K. Potter researched bifurcation of the saddle shape of a [0/90] composite [16]. During this study, the researchers observed the saddle shape was unstable compared to the other two cylindrical shapes. Using a numerical models, they compared the Rayleigh-Ritz method model to the numerical models.

As seen in the literature, bi-stability of the composite laminate is achieved during fabrication. After the cure process, multiple deformation shapes can be observed based on the ply orientation and material [9]. This deformation is due to the thermal strain gradient between the layers of the laminate and cure shrinkage during the cure process [10]. As an essential part of the smart material morphing system, the cure process and the resulting post-cure deformation shapes are heavily investigated. Thin unsymmetric composite laminates will deform to one of two circular cylindrical shapes, with the shape in between these two cylindrical shapes being unstable and called the saddle shape. Work has been done to show that the CLT can only predict that the post-cure room temperature shape of all thin composite laminates will be the saddle shape [17]. A Rayleigh-Ritz technique has been established and used here to study the post-cure deformation shapes. This paper presents the CLT first then builds upon it with the methods of strain energy and virtual work for approximations used by M. W. Hyer, Jones, and others to get better post-cure room temperature shape predictions. During the cure process, there are certain mechanisms which cause the displacement and shape change of the composite [18]. With these assumptions, the CLT is prone to error when determining the post-cure room temperature shape. The numerical models have the capabilities to couple thermal, electrical, and mechanical domains and allows user defined equation integration. This phase of research entails modeling the cure process, which is used to more accurately characterize the piezoelectric attachment and actuation in the numerical models. To validate the techniques and methods used, the analytical and numerical models are compared to experimental results.

2. ANALYTICALLY MODELING THE CURE PROCESS

The fabrication and cure process of unsymmetric laminates are essential elements of the smart material morphing structure. Models were developed to achieve a better understanding of these processes for piezoelectric actuation. An analytical model is presented first to predict the post-cure shape and deformation. The analytical model was based on both Hyer's and Vizzini's composite material work [3,19,20]. Numerical analysis was done to compare to the analytical results, and is presented following the analytical models. Multi-physics Finite Element Models were created for numerical modeling of both the fabrication process and piezoelectric actuation [21]. Experimental setup and procedures for fabrication and the cure process are then explained. Analytical, numerical, and experimental results will be shown. A post-cure shape deformation comparison between the models will be shown in later sections.

2.1. Classical Lamination Theory

AS4 carbon fiber pre-impregnated laminae are used for the composite laminates of this study. Due to the thickness of the carbon fiber pre-pregs, the laminates in this study are classified as thin composite laminates where the thickness to length ratio is less than $\frac{1}{20}$. This ratio is important for post-cure deformation shapes, thin laminates will deform to a circular cylindrical shape whereas a thicker laminate will deform to that of the saddle shape. The first model was done with the CLT for composite laminates. For CLT analysis, the following key assumptions of the material are made:

- assume smeared properties through thickness;
- assume perfectly bonded laminae with no defects;
- assume zero bond line thickness between plies (no epoxy between plies);
- assume Kirchhoffs hypothesis is valid; plane sections remain planar and perpendicular sections remain perpendicular.

The first assumption states that the individual fibers of the laminae are homogeneous, not varying with direction. The second and third assumptions are made to eliminate any complications with bonding the laminate. The models of the laminate are assumed to be perfectly bonded with no epoxy layer between the plies. The fourth assumption is made in relation to Kirchhoffs hypothesis. This assumptions states that the strain is linear through the thickness of the laminate. The laminate is then treated as an orthotropic material with three planes x, y, and z (also known as 1, 2, and 3). Treating the laminate as an orthotropic material, nine engineering constants of the material are required including Youngs modulus, Poisson ratio, and shear modulus of the material one in each directional plane. These constants are then used to determine the reduced material stiffnesses of the laminate [20]. In order to observe how material stiffness relates to each individual lamina with a specified orientation, the stressstrain relations need to be transformed. Using a method on invariants, the lamina stress-strain relations can be determined for an arbitrary orientation. In terms of the reduced stiffnesses, these invariants can be calculated

$$I_1 = \frac{Q_{11} + Q_{22} + 2Q_{12}}{4} \tag{1}$$

$$I_2 = \frac{Q_{11} + Q_{22} - 2Q_{12} + 4Q_{66}}{8} \tag{2}$$

$$R_1 = \frac{Q_{11} - Q_{12}}{2} \tag{3}$$

$$R_2 = \frac{Q_{11} + Q_{22} - 2Q_{12} - 4Q_{66}}{8} \tag{4}$$

where Q_{ij} are the reduced material stiffnesses [9]. Using these invariants, the transformed reduced stiffness tensor can now be found:

$$E_{11}(\theta) = I_1 + I_2 + R_1 \cos(2\theta) + R_2 \cos(4\theta) \tag{5}$$

$$E_{22}(\theta) = I_1 + I_2 - R_1 \cos(2\theta) + R_2 \cos(4\theta) \tag{6}$$

$$E_{12}(\theta) = I_1 - I_2 - R_2 \cos(4\theta) \tag{7}$$

$$E_{16}(\theta) = \frac{1}{2} R_1 \sin(2\theta) - R_2 \sin(4\theta) \tag{8}$$

$$E_{66}(\theta) = I_2 - R_2 \cos(4\theta) \tag{9}$$

$$E_{26}(\theta) = \frac{1}{2} R_1 \sin(2\theta) + R_2 \sin(4\theta) \tag{10}$$

where θ is the arbitrary ply orientation. The next step in the analysis is to calculate the forces and moments that occur during the cure process. In order to determine these resultant forces and moments the extensional, bending-extension coupling, and the bending stiffness tensors also named the ABD matrix, must be calculated. Using the rotated stiffness tensor equations and the middle surface lamina thickness, the ABD matrix can be determined for each direction of the laminate [9]. The B tensor component of the ABD matrix, the bending-extension coupling stiffness, implies that if the composite is pulled a bending and/or twisting reaction of the composite must occur also. For the thin cross-ply unsymmetric composites, this bend/twist coupling is not expected to occur. Treating the laminate as an orthotropic material comes from the the D_{ij} component of the ABD matrix which is the bending stiffness of the laminate. If $D_{16} = D_{26} = 0$, then this assumption can be made. Unsymmetric laminates contain this type of coupling stiffness while symmetric laminates do not. The force and moment constitutive equations are given by

$$\begin{Bmatrix} N_{11} \\ N_{22} \\ N_{12} \\ M_{11} \\ M_{22} \\ M_{12} \end{Bmatrix} = \begin{bmatrix} A_{11} & A_{12} & A_{16} & B_{11} & B_{12} & B_{16} \\ A_{12} & A_{22} & A_{26} & B_{12} & B_{22} & B_{26} \\ A_{16} & A_{26} & A_{66} & B_{16} & B_{26} & B_{66} \\ B_{11} & B_{12} & B_{16} & D_{11} & D_{12} & D_{16} \\ B_{12} & B_{22} & B_{26} & D_{12} & D_{22} & D_{26} \\ B_{16} & B_{26} & B_{66} & D_{16} & D_{26} & D_{66} \end{bmatrix} \begin{Bmatrix} \varepsilon_{11}^{o} \\ \varepsilon_{22}^{o} \\ \gamma_{12}^{o} \\ \kappa_{11}^{o} \\ \kappa_{22}^{o} \\ \kappa_{12}^{o} \end{Bmatrix} \tag{11}$$

where N and M are the forces and moments. To calculate these strains and curvatures values, the ABD matrix can be inverted to solve for the desired variables. If the forces and moments were to be determined and the strains and curvatures were known, the simple calculation can be done with the matrices. Continuing the CLT, the laminate is treated as orthotropic material to equate the thermal strains to the coefficients of thermal expansion multiplied by the thermal gradient. The residual stresses and strains arising from the cure process can then be found [20].

2.2. Rayleigh-Ritz Technique

The CLT predicts a saddle shape will occur each time a thin unsymmetric laminate cools from the elevated cure temperature to room temperature. This is due to the failure to capture reference strain due to the large out-ofplane deformations as the laminate is cooled. The Rayleigh-Ritz technique accounts for the large out-of-plane deformations arising from the room temperature cooling. The maximum states of the potential energy are the unstable equilibrium conditions of the laminate. The minimum states of the potential energy are the conditions needed to characterize the deformations of the laminate cooling to room temperature. Using the same engineering constants, mentioned earlier, the reduced stiffnesses of the laminate are found. The coefficients of thermal expansion are also transformed to the ply orientation not aligned in the principal axis. The process is continued to using the method of invariants mentioned in the earlier section to determine the reduced stiffnesses of the laminate. The thermal strains and stresses are then evaluated using the following equations:

$$\sigma_x^T = Q_{11}\varepsilon_x^T + Q_{12}\varepsilon_y^T + Q_{16}\gamma_{xy}^T \tag{12}$$

$$\sigma_y^T = Q_{12}\varepsilon_x^T + Q_{22}\varepsilon_y^T + Q_{26}\gamma_{xy}^T \tag{13}$$

$$\sigma_{xy}^T = Q_{16}\varepsilon_x^T + Q_{26}\varepsilon_x^T + Q_{66}\varepsilon_x^T \tag{14}$$

where the thermal strains, ε^T, are given by:

$$\varepsilon_x^T = \alpha_x \Delta T \tag{15}$$

$$\varepsilon_y^T = \alpha_y \Delta T \quad (16)$$

$$\gamma_{xy}^T = \alpha_z \Delta T \quad (17)$$

where αi are the coefficients of thermal expansion. The middle surface strains and curvatures can then be approximated. Where u°, v°, and w° are the displacement fields:

$$u^o = cx - \frac{a^2x^3}{6} - \frac{abxy^2}{4} \quad (18)$$

$$v^o = dy - \frac{a^2y^3}{6} - \frac{abx^2y}{4} \quad (19)$$

$$w^o = \frac{1}{2}\left(ax^2 + by^2\right) \quad (20)$$

The stresses and strains are then approximated using

$$\varepsilon_x = \varepsilon_x^o + z\kappa_x^o \quad (21)$$

$$\varepsilon_y = \varepsilon_y^o + z\kappa_y^o \tag{22}$$

$$\gamma_{xy} = \gamma_{xy}^o + z\kappa_{xy}^o \tag{23}$$

$$\sigma_x = Q_{11}\varepsilon_x + Q_{12}\varepsilon_x + Q_{16}\gamma_{xy} - \sigma_x^T \tag{24}$$

$$\sigma_y = Q_{12}\varepsilon_x + Q_{22}\varepsilon_y + Q_{26}\gamma_{xy} + \sigma_y^T \tag{25}$$

$$\sigma_{xy} = Q_{16}\varepsilon_x + Q_{26}\varepsilon_y + Q_{66}\gamma_{xy} - \sigma_{xy}^T \tag{26}$$

The post-cure room temperature shape can then be modeled. This step does not include bonding the MFC actuator, therefore only thermal effects are accounted for. The strain energy or total potential energy is given by:

$$\Pi_1 = \frac{1}{2}\int_{\frac{-Lx}{2}}^{\frac{Lx}{2}}\int_{\frac{-Lx}{2}}^{\frac{Lx}{2}}\int_{z_2}^{z_o}\Big[\left(\sigma_x - \sigma_x^T\right)\varepsilon_x + \left(\sigma_y - \sigma_y^T\right)\varepsilon_y + \left(\sigma_{xy} - \sigma_{xy}^T\right)\gamma_{xy}\Big]\mathrm{dxdydz} \tag{27}$$

Carrying out the integrations in equation 27, the potential energy is then reduced to an algebraic equation in terms of the coefficients a, b, c, and d. Once the integrations are computed, Equation (27) will show that the potential energy is also in terms of material properties and geometries of the laminate. These coefficients are then determined by solving the nonlinear algebraic equations that is reduced from equating to zero the first variation of the potential energy with respect to the coefficients,

$$\frac{\partial \Pi}{\partial C_i} = 0, i = 1, \cdots, 4 \tag{28}$$

where C is a, b, c, and d. The undetermined coefficients were found using Newton's iterative method. These solutions relate to the equilibrium shapes of the cooled room temperature laminate. To check stability of the solution for the laminate, the second variation of the potential energy must be positive definite.

The following is an example of the results found for the analytical methods mentioned for the post-cure room temperature shapes of the laminates. Table 1 shows the material properties of the carbon fiber used.

The laminates studied were [0/90], [0_2/90_2], [0/45], and [45/90] square and rectangular laminates. When cooling to room temperature, the cross-ply laminates will deform to one of three possible equilibrium shapes. Two being stable cylindrical shapes and one being the unstable saddle shape, the shape between the two cylindrical shapes. After analytical formulation, these equilibrium shapes can be predicted and are shown in the meshes of Figure 1 for the [0/90] laminate.

Table 1. Material properties of AS4 carbon fiber.

AS4/3501 Carbon Fiber	
E_1	150 GPa
E_2	20 GPa
v_{12}	0.25
a_1	$-7.815e^{-10}$ 1/K
a_2	$6.252e^{-8}$ 1/K
G_{12}	5 GPa
Curie Temperature	449.817 K

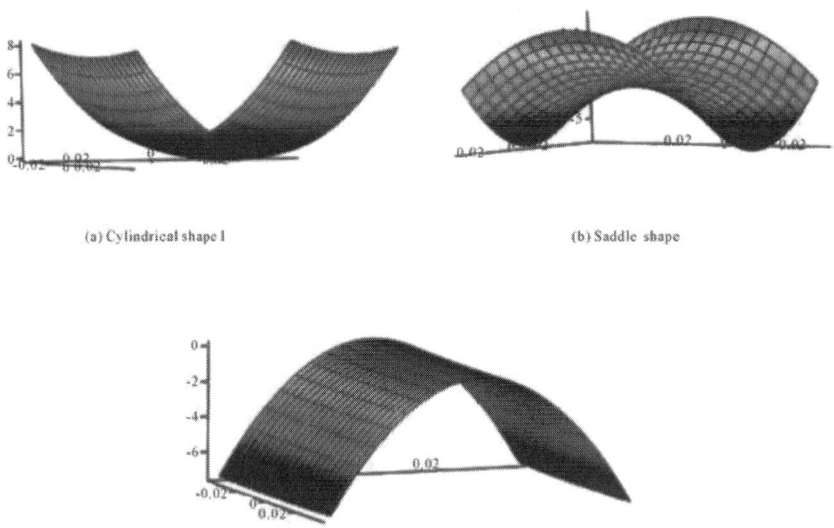

(a) Cylindrical shape I

(b) Saddle shape

(c) Cylindrical shape II

Figure 1. [0/90] Cooled room temperature shapes.

As seen in plot two of Figure 1, the saddle shape can be referred to as the shape between the two cylindrical shapes. For the simple cross-ply laminates, this shape should not occur after cooling to room temperature. The different orthogonal cylindrical curvatures can be seen and compared in plots one and three. For the [0/90] laminate, the expected post-cure deformation shape is the shape in plot one. After piezoelectric actuation, the [0/90] laminate should snap through to the second cylindrical shape two, shown in plot three of Figure 1. The transverse displacement after the curing process was also studied. This was done to investigate the curvatures of both the post-cure shape and the later piezoelectric actuated shapes. For the first cylindrical shape, a (expected shape after curing), the transverse curvature is along side length Lx. This is the curvature for the ideal situation. The third plot shows the second room temperature cylindrical shape of the [0/90] laminate.

Next, predictions can be made for the [0_2/90_2] laminate using the same techniques as described earlier. With more plies than the [0/90] laminate, the curvature is expected to be different due to the increase of plies. The transverse curvature for the first cylindrical shape (expected shape after curing) of the [0_2/90_2] laminate is expected to be along the length of side L_y. When compared with the [0/90] laminate this curvature is in opposite the direction along the other axis. The equilibrium shapes, shown as meshes, of the [0_2/90_2] laminate in Figure 2 confirms the different post-cure shape of the laminates.

The first plot (a) in Figure 2 is the expected post-cure shape of the [0_2/90_2] laminate. Plot b represents the shape between the two cylindrical shapes, the saddle shape. Plot c shows the expected actuated displacement shape of the [0_2/90_2] laminate.

The [0/45] laminate was modeled using the RayleighRitz technique as described above. With a ply orientation at 45°, it is expected that the laminate will not deform to one of the equilibrium cylindrical stable shapes. Instead, the [0/45] laminate should deform to the saddle shape. As research shows, this shape is unstable compared to the simple cross-ply laminates. The stability is determined by the second variation of the total potential energy, or the second derivative of the Jacobian matrix. If this matrix is positive definite, which confirms that the equilibrium solution is stable. As shown in Figure 3, the analytical models predicted the post-cure shape of the [0/45] laminate will deform to the saddle shape. The meshes in Figure 3 may seem flat when compared to the simple cross-ply laminates, but that is due to the instability of the laminate.

The curvature of the [0/45] laminate is along both the x and y axis. This is a characteristic of the saddle shape. After fabrication, the [0/45] laminate should deform to this saddle shape. For piezoelectric actuation the [0/45] laminate is still going to snap through to another saddle shape (shown in later chapters), but unlike the simple cross-ply laminates, due to the instability of the [0/45] laminate removal of the snap through force will cause the laminate to snap back to its original shape shown in Figure 3. These analytical results can now be used as predictions of the post-cure room temperature shapes as well as models to help accurately capture the effects of the piezoelectric actuation models. Comparisons were made to the numerical models and the experimental results.

3. NUMERICAL MODELING THE CURE PROCESS

Multi-physics numerical FEA models were used to compare analytical models and experimental results. The present models required implementing the Thermal-Structural Interaction mode, which couples both, the structural mechanics and the heat transfer modules. The analysis type of the stress-strain and the heat transfer modules were chosen as static and transient types respectively. Due to the small thicknesses of the laminates studied, models were done using linear and quadratic mesh element types.

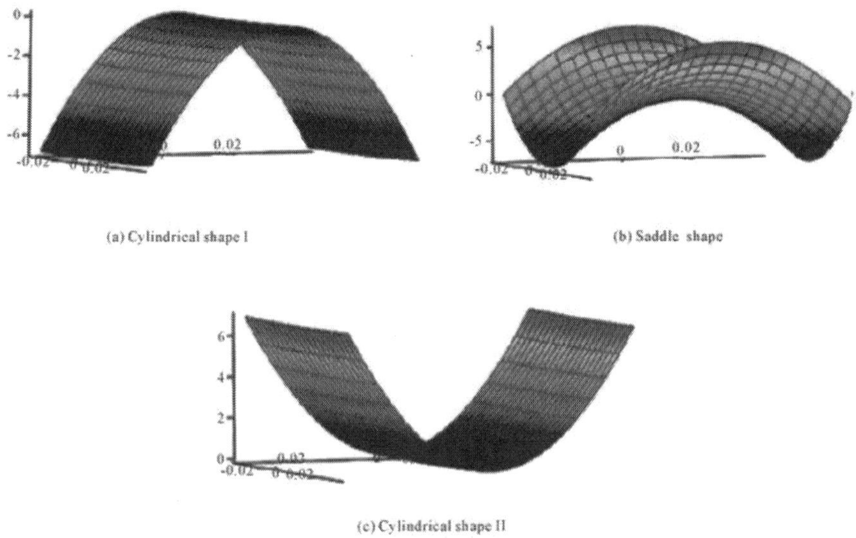

(a) Cylindrical shape I

(b) Saddle shape

(c) Cylindrical shape II

Figure 2. $[0_2/90_2]$ Cooled room temperature shapes.

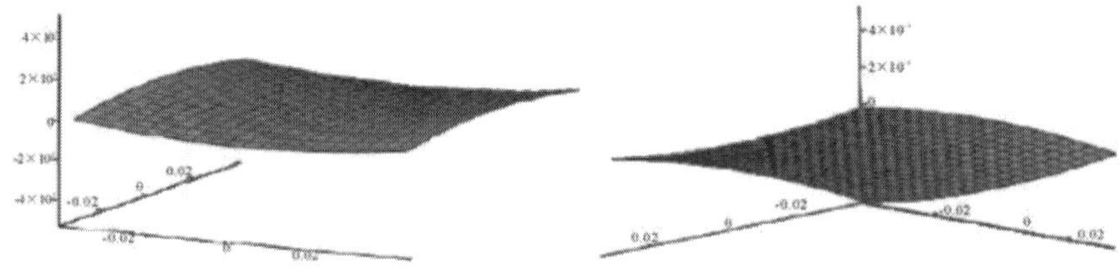

Figure 3. [0/45] Cooled room temperature shapes.

3.1. Initial Setup

The Thermal-Structural interaction mode allowed for including thermal expansion due to the coefficients of thermal expansion(CTE) during the cure process. The CTE's along with geometry and other material parameters, play key roles during the heating and cooling of the laminate. In numerical model 3-D space, new coordinate systems are needed to specify certain fiber orientations for plies not oriented in the global or principal material axes. This feature is done by rotating the global coordinate system from the original x, y, z axis to the angle needed for the specific ply. Allowing 0° to represent the default or global coordinate system (no need to change coordinate system if ply is oriented at 0°), the remaining plies coordinate systems can be changed using the consecutive rotation axes option. After the laminate is drawn with the specified dimensions, the material properties and conditions can be specified in the structural mechanics subdomain settings of the model. The heat transfer sub-domain settings controls the heat source and the thermal properties of the laminate. The structural domain settings control the material properties of the laminate, including the fiber orientation.

The laminate model was treated as an orthotropic material as stated earlier. In these settings, the necessary engineering constants along with the selection of the global coordinate system and specified created coordinate systems can be entered. The initial pressure of the laminate can be selected. An initial pressure was used to model the mold pressing the laminate flat during the heating of the laminate. Similar to the sub-domain settings, there are separate boundary condition settings for the structural application and the heat transfer application. Modeling free thermal expansion, there were no constraints on the boundaries or edges of the laminate. Compared with the analytical model, this assumption can be made due to the absence of external loads during the cooling of the laminate to room temperature. When solving the model, the solver parameters are important for accuracy and solver convergence. For the structural application mode, the solver was chosen based on the static analysis nature of the problem. This analysis type differs from the heat transfer application mode, however. For the heat transfer application mode, the analysis was chosen to be transient based on the steps of the curing process. Therefore, the analysis type needed to couple both the static and transient nature of the curing process.

3.2. Coupling Structural Mechanics and Heat Transfer

The numerical analysis allowed for more complex lamina behavior than the CLT assumptions used in the analytical modeling. For the structural mechanic analysis mode of the numerical modeling, a few important equations were used to define the domain of the laminate. Using the solid, stress-strain application for 3-D modeling, the straindisplacement relationship equations were specified as follows:

$$\varepsilon_x = \frac{\partial u}{\mathrm{d}x} \tag{29}$$

$$\varepsilon_y = \frac{\partial v}{\partial y} \tag{30}$$

$$\varepsilon_z = \frac{\partial w}{\partial z} \tag{31}$$

$$\varepsilon_{xy} = \frac{1}{2}\left(\frac{\partial u}{\partial y} + \frac{\partial u}{\partial x}\right) \tag{32}$$

$$\varepsilon_{xy} = \frac{1}{2}\left(\frac{\partial v}{\partial z} + \frac{\partial w}{\partial y}\right) \tag{33}$$

$$\varepsilon_{xz} = \frac{1}{2}\left(\frac{\partial u}{\partial z} + \frac{\partial w}{\partial x}\right) \tag{34}$$

The stress-strain relation of the numerical analysis is given by the generalized Hooke's Law formulation:

$$\sigma = D\varepsilon \tag{35}$$

where D is the stiffness matrix of the material. The Structural Mechanics mode is based on the weak formulation of equilibrium equations in the global stress components. For 3-D models, the equilibrium equation is represented by:

$$-\nabla\sigma = F \tag{36}$$

where σ is the stress tensor and F is the volume or body forces [21].

The sub-domain setting for the heat transfer mode was specified using Fourier's law:

$$-\nabla\left(k\nabla T\right) = Q \tag{37}$$

where T is the temperature being measured at that instant, Q is the heat source, and k is the thermal conductivity. Specifying the heat transfer boundary equation using the heat flux equation,

$$n\left(k\nabla T\right) = q_o + h\left(T_{\text{inf}} - T\right) + \text{Const}\left(T_{amb}^4 - T^4\right) \tag{38}$$

where q_o is the inward heat flux of the composite being modeled, h is the heat transfer coefficient, T_{inf} is the external temperature, Const is the numerical constant for heat transfer relations, and T_{amb} is the ambient temperature. The heat transfer boundary equation is a form of the heat convection equation. The second half of the heat transfer boundary equation is the heat transfer due to radiation. Although shown, it was not used in the models. The equations in this section are all derived or are a form of the same analytical equations used in the numerical analysis. Using the same material properties as in Table 1, the laminates were modeled and the post-cure room temperature shapes were predicted.

Applying an initial pressure of 5.861×10^5 Pa to the laminate models the press applied from the mold during the heating of the laminate. The laminate is then heated to 449.817 K then cooled to room temperature, 296.483 K. Figure 4 shows the numerical solutions of the postcure room temperature shape for the [0/90] rectangular and square laminates. It is seen that for both [0/90] laminate geometries, the symmetric circular transverse curvature is in the negative z-direction. This curvature can be seen in the first plot of Figure 4 along the x-direction and along the y-direction in the second plot.

The [0/90] laminate was modeled with a total thickness of 0.36002 mm with square length of 76.2 mm and rectangular lengths of L_x = 61 mm and L_y = 120 mm which classified the laminates as thin laminates. In addition to being thin laminates, they were simple cross-ply laminates therefore the saddle shape is not present here. Increasing the number of plies to a $[0_2/90_2]$ laminate, different cylindrical room temperature shapes are observed. Figure 5 is the numerical solution of the room temperature shape of the $[0_2/90_2]$ laminate. The $[0_2/90_2]$ laminate was modeled at 0.72004 mm total thickness. The lengths were the same as the [0/90] laminate. Therefore the saddle shape is not seen here. For the $[0_2/90_2]$ laminate, the symmetric circular transverse curvature is in the positive z-direction. This curvature is seen in the first and second plot of Figure 5 along the y-direction.

For laminates that are not simple cross-ply laminates, a saddle shape as the room temperature post-cure shape is expected. Experimental and analytical work [6] has been done to better understand the saddle shape and the instability of the laminate. The simple cross-ply laminates deform while cooling to room temperature to a symmetric circular shape. This is seen in the numerical solutions above. More complex laminates will deform unsymmetrically with a twist curvature similar to the saddle shape. This twist curvature develops as the laminate cools to room temperature. Applying a force to this saddle shape will cause deformation, but unlike the simple cross-ply laminates, removal of the snap through force will cause the laminate to return to its original shape. This classifies the laminate in the saddle shape as unstable. For this study, [0/45] and [45/90] were used for experimentation. Using the same parameters as the simple cross-ply laminates, the [0/45] laminates were modeled.

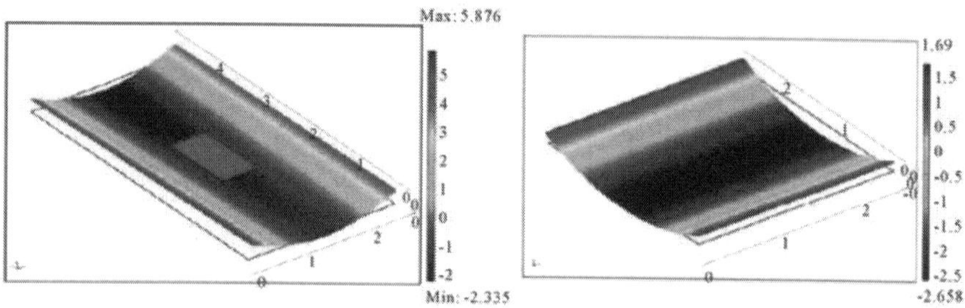

Figure 4. Rectangular and square [0/90] cylindrical shape I.

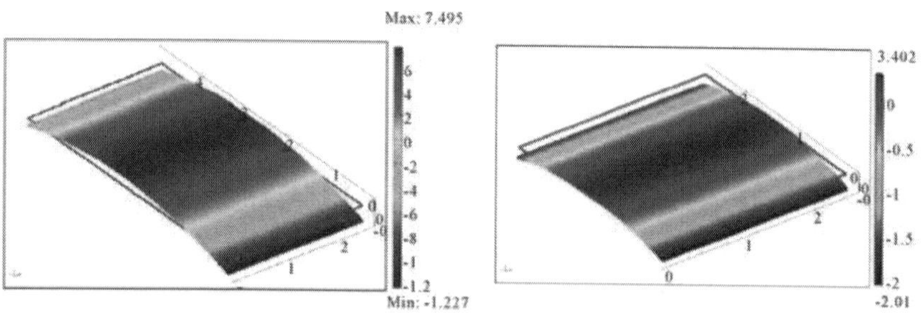

Figure 5. Rectangular and square [0_2/90_2] cylindrical shape I.

Figure 6 shows the rectangular and square [0/45] laminate at room temperature. The saddle shape is observed due to the 45° ply. The twist curvature which characterizes the saddle shape is seen in all three plots. For the rectangular [0/45] laminate in plot 1, two edges deform upward in the positive z-direction while the other two edges deform downward in the negative z-direction.

4. EXPERIMENTAL FABRICATION AND RESULTS

4.1. Experimental Fabrication

A standard hand lay-up procedure was used to stack the plies of the laminate. Both square and rectangular laminates were fabricated to better understand the role of the material properties and geometry in piezoelectric actuation. A mold was designed to press the laminate flat while heating to the cure temperature. A convection oven at Mississippi State University Raspet Flight Research Laboratory was used for curing the composite laminates. The mold held the composite laminate flat while in the oven at a cure temperature of 449.817 K. Unsymmetric laminates and the snap through phenomenon work best when the laminate can be classified as a thin laminate.

After removal of the mold from the oven, the press is released allowing the laminate to deform freely. Thin laminates, where the length and width are large compared to the thickness, will deform to one of two cylindrical shapes as discussed earlier [8]. The laminate is then left over night to complete cooling. Figure 7 shows both the [0/90] and $[0_2/90_2]$ laminate after the cure process. The circular curvature seen in Figure 7 for the [0/90] laminate and the $[0_2/90_2]$ is similar to the prediction of the analytical and numerical models.

The post-cure transverse curvature in the $[0_2/90_2]$ is greater than the curvature of the [0/90] laminate. This is possible due to the increase of the number of plies in the laminate. Although the transverse curvature is greater, with more plies the laminate becomes stiffer, therefore making it more difficult to apply force for snap through without damaging the laminate.

The next composites fabricated were the laminates that were not classified as cross-ply laminate. These laminates, as discussed earlier, is expected to deform to a saddle shape. The twist curvature that arose from the cooling of the laminate is seen in both the [0/45] and [45/90] laminates in Figure 8. The curvature is similar to numerical models presented earlier.

For comparisons seen in later sections, the transverse displacement of the deformed laminates were measured. Due to the symmetric curvature of the simple cross-ply laminates, $[0_n/90_n]$ where n is the number of plies, this symmetric cylindrical shape can be measured using a FARO gage Coordinate-measuring machine (CMM). The CMM establishes a coordinate system for the laminate. This coordinate system is then used to measure the deformation and mesh the post-cure room temperature laminate. The difference between the post-cure room temperature symmetries of the cross-ply laminates and other laminates can now visibly be seen. For the cross-ply laminates, the post-cure room temperature curvatures were symmetric as expected. For the [0/45] and [45/90] laminates, the curvature is not symmetric as the cross-ply laminates. This was expected and predicted due to the ply orientation of the laminates. This saddle shape when compared to the cylindrical shapes of the cross-ply laminates is classified as an unstable shape. Applying a force to the post-cure room temperature shapes causes the laminates to snap through to another shape. For the crossply laminates this shape is the second cylindrical shape seen in previous sections. After snap through, this shape remains upon removal of the snap through force, therefore being a stable shape. For the laminates not classified as cross-ply laminates, the laminate will snap back to its original (post-cure room temperature) shape upon removal of the snap through force. Therefore these laminates are unstable.

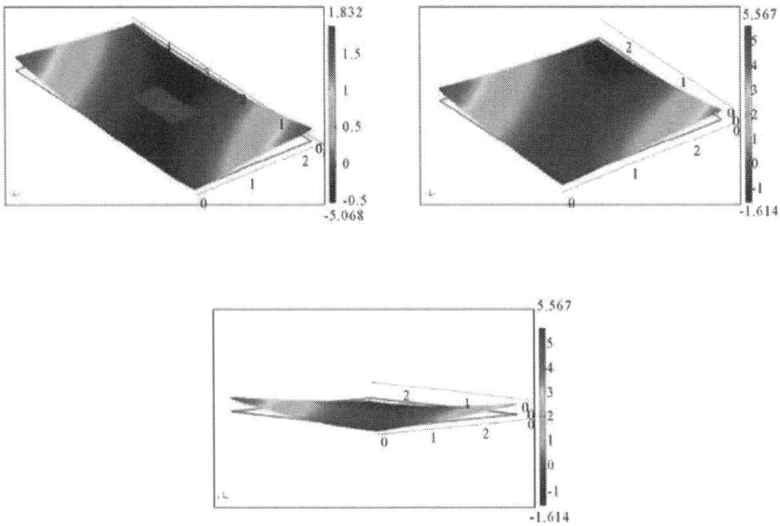

Figure 6. Rectangular and square [0/45] shape I.

Figure 7. Cured room temperature shapes of [0/90] and $[0_2/90_2]$.

Figure 8. Cured room temperature shapes of [0/45] and [45/90].

4.2. Experimental Comparisons

For the analytical method, the Rayleigh-Ritz technique along with the total potential energy was used to predict the post-cure shape of the thin laminates. After fabrication, the profile of the laminates were measured using the CMM gage instrument. This instrument provided a coordinate system and data points for the deformed laminate, which provided meshes of the laminates. The data points can also be compared to the transverse post-cure curvatures of the analytical and numerical models. Because laminates wet thin and the CMM required contact, laminates were easily moved vertically by the gage arm of the CMM. When collecting the data points of the post-cure shape, the CMM arm would forcefully displace the laminate up or down. This caused a few errors when plotting the data points of the laminates from the CMM instrument. Laminates were taped to the measurement table to increase stiffness and minimize deflection to the the CMM. Although this provided better control of the arm and laminates, the error was still evident as the CMM arm moved across the laminate.

The post-cure fabrication results of the [0/90] rectangular laminate is shown in Figure 9 below.

The comparisons for the experimental results and the numerical model are shown in Figure 9. The solid black line represents the experimental results and the red traingles represent the numerical model. In plot one, the curvature is given along the x-axis from end to end. The coordinate system here places the origin at the edge of the laminate. Although reasonably close, the error seen is due to the probe of the FARO CMM arm. The same results can be shown for the square [0/90] laminate in Figure 10. As expected, the post-cure curvatures for the [0/90] laminate are in the negative z-direction along the x-axis. This post-cure curvature is expected to be opposite for the $[0_2/90_2]$ laminates.

The $[0_2/90_2]$ laminates have been compared similar to the [0/90] laminates. As seen in other sections, the postcure curvature of the $[0_2/90_2]$ laminates are in the positive z-direction along the y-axis. The different curvature of the $[0_2/90_2]$ and the [0/90] laminate is due to the more dominate 90° plies. While fabricating the laminates, it was observed after cutting to size, that the 90° carbon pre-preg would start to deform to this shape when left alone. Therefore, during cooling of the $[0_2/90_2]$ laminates this deformation is to be expected. Similar to the [0/90] laminate, the numerical model will be shown as red triangles on the plot while the experimental results will be represented by solid black lines in Figure 11.

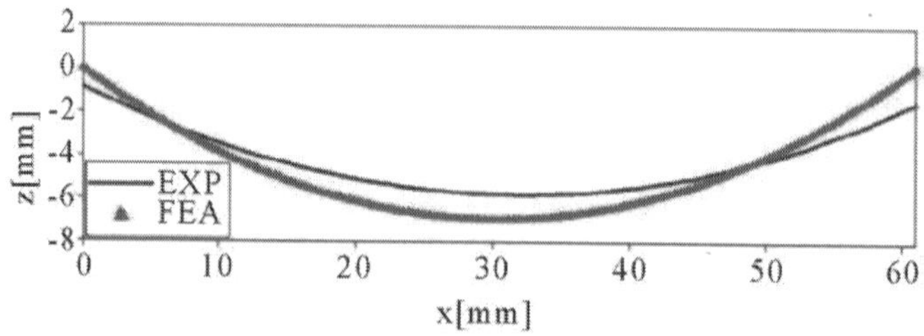

Figure 9. Rectangular [0/90] post-cure room temperature displacement.

The maximum post-cure curvatures for the simple cross-ply laminates are shown in Table 2. The laminates not classified as the simple cross-ply laminates could not be measured with the FARO CMM gage due to the unsymmetric post-cure curvature. Therefore the other laminates were investigated by the use of numerical models and experimentation. The analytical and numerical models can now serve as preliminary models necessary for piezoelectric actuation.

The comparisons of the [0_2/90_2] laminate for the numerical model and the experimental results are shown in plot one of Figure 11. The curvature shows that the maximum magnitude for the post-cure transverse displacement is going to occur at the center laminate, approximately by 7.5 mm. The same comparisons are made for the square [0_2/90_2] laminate.

5. PIEZOELECTRIC ACTUATION

Using the results of the sequential models of the fabrication process, the effects of piezoelectric actuation were captured. These preliminary models enhance the knowledge of the key parameters needed for piezoelectric actuation. The MFC actuator used for experimentation is distributed by Smart Material Corporation. The MFC was invented by NASA in 1996 and since then has been continued to be improved for many applications. The piezoelectric material of the MFC actuator are r piezoceramic rods of the actuator [1]. These rods are sandwiched between electrodes and polyimide film as shown in Figure 14. The electrodes are in an interdigitated pattern allowing for direct application of the a voltage to the rods, which is the converse piezoelectric effect. Once the voltage is applied, the actuator will elongate or deform. This system enables in-plane poling and actuation, both which are used for this research. The operation mode used for experimentation was the d_{33} operating mode. This mode elongates the rods when a voltage is applied. The tensor notation for the operation mode can be broken down to better understand the proper applications and differences between other operation modes. The first variable, 3, represents the axis of polarization for the material. This direction 3, is through the thickness of the material also known as the z-direction if in x, y, z coordinates. Making the 1 and 2 directions align in the plane of the piezoelectric material. The second variable, 3, represents the direction of which the state of the piezoelectric material should be analyzed [22]. With the d_{33} operation mode and the assumption of free strain, the constitutive equations are reduced. Free strain is the strain produced in the piezoelectric material when there is no resistive stress on the material. Piezoelectric actuation will be shown for the [0/90] laminates.

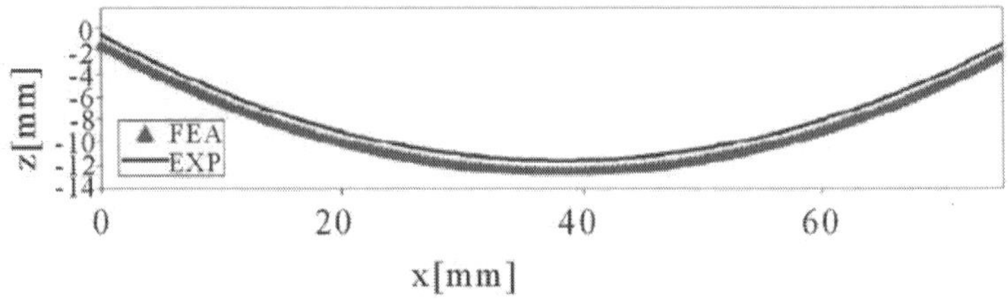

Figure 10. Square [0/90] post-cure room temperature displacement.

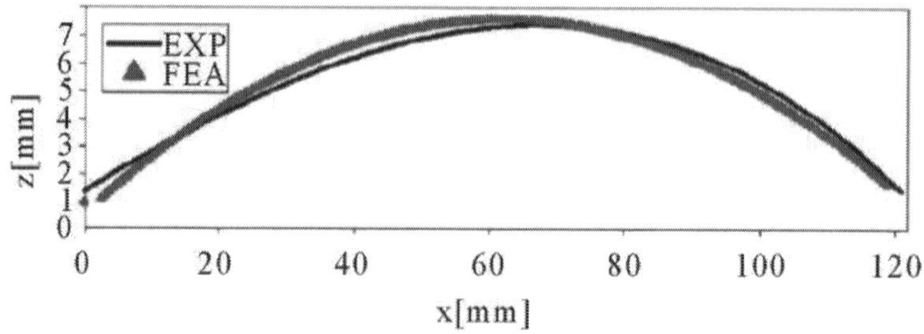

Figure 11. Rectangular [0_2/90_2] post-cure room temperature displacement.

Table 2. Maximum post-cure displacement curvatures (mm).

	[0/90] rect.	[0/90] square	[0_2/90_2] rect.	[0_2/90_2] square
Analytical	7.177	9.232	6.0148	7.477
Numerical	6.342	10.632	6.02	8.21
Experimental	5.649	10.711	5.716	6.552

5.1. Analytical Morphing Models

For piezoelectric analysis, the plates or in this case MFC actuators, were classified as thin plates. Thin plates are those with a ratio of thickness to length less than 1/20. This allows the complex three-dimensional problem to be reduced to a problem in two dimensions [22]. Using the same assumptions of the classical plate theory, the thin plate is now classified as a plane stress problem. The analysis implements the total potential energy approach similar to the post-cure shape analytical work. The Ritz method is used for the variational principal solution and is reduced to a set of linearly independent equations by an assumed form of generalized coordinates. This same procedure was done in previous sections for modeling the cure process. The post-cure room temperature shapes in previous sections had to be accounted for in piezoelectric analysis. When using the total potential energy to model the post-cure shapes, the driving force was the heat transfer taking place. Now for piezoelectric analysis, the driving force will be piezoelectricity. The total potential energy in response to the MFC actuation is given by:

$$\Pi_2 = \frac{1}{2} \int_{-\frac{LxMFC}{2}}^{\frac{LxMFC}{2}} \int_{-\frac{LyMFC}{2}}^{\frac{LyMFC}{2}} \int_{z_2}^{z_3} \Big[\left(\sigma_x^a - \sigma_x^{Es}\right)\varepsilon_x + \left(\sigma_y^a - \sigma_y^{Es}\right)\varepsilon_y + \left(\sigma_{xy}^a - \sigma_{xy}^{Es}\right)\gamma_{xy} \Big] \mathrm{d}x\mathrm{d}y\mathrm{d}z \tag{39}$$

where $\sigma_x^a, \sigma_y^a, \sigma_{xy}^a, \sigma_x^{Es}, \sigma_y^{Es}, \sigma_{xy}^{Es}$ are the stresses and

$\varepsilon_x, \varepsilon_y$ and γ_{xy} are the strain strains in the MFC actuator which include the stresses induced by bonding the MFC actuator to the laminate [23]. The total potential energy for the actuation of the laminate, taking the postcure deformation into account is then given as:

$$\Pi_3 = \Pi_2 + \Pi_1 \tag{40}$$

Using the same techniques, the undetermined coefficients a, b, c, and d can be found by minimizing the total potential energy. Thus giving the actuated shapes for the MFC actuator and laminate.

For the post-cure [0/90] laminate, the curvature was in the negative z-direction along the x axis. For the cylindrical shape II, the curvature is expected to be in the positive z-direction along the y axis (this would be known as the orthogonal cylindrical shape). The postcure shape (cylindrical shape I) of the $[0_2/90_2]$ laminate has transverse curvature in the positive z-direction along the yaxis. The orthogonal cylindrical shape to the deformed post-cure shape of the $[0_2/90_2]$ laminate would have curvature in the negative z-direction along the x-axis.

As shown in Figure 12, the actuated deformation obtained from the total potential energy method predicted the [0/90] laminate to deform in the positive z-direction. The expected curvature for the $[0_2/90_2]$ laminate has also occurred.

5.2. Actuation Modeling

After the curing process is simulated, the same models can be used to model the piezoelectric actuation. Running the structural mechanics and heat transfer modules together, gave the post-cure room temperature shapes of the cross-ply laminates. Using these existing models, the 3D piezoelectric solid module was added. This allowed for the same post-cure curvature (same geometries from cure process models) to be used for modeling piezoelectric actuation. The same element types were used as in the fabrication models. After drawing the rectangular 27.9 × 13.995 mm MFC actuator at the center of the laminate on the first ply, the laminate and actuator was bonded to the laminate by creating a contact pair between the bottom surface of the MFC actuator and the top surface of the laminate. This was done to make the MFC actuator and laminate active in one domain. The actual MFC actuator has piezoelectric fibers along the length of the actuator, which corresponds to the zero direction. Therefore, for the numerical models, the MFC actuator was bonded to the laminate according to the direction of the first ply. Using the same structural mechanics domain settings from the cure process models, the ply orientation, thickness, and material are recalled for the 3D piezoelectric solid module. These models implement applying the voltage to the MFC actuator using the following piezoelectric equations in the stress-strain form.

$$T = c^E S - \mathrm{e}^T E \tag{41}$$

$$D = \mathrm{e}^S + \varepsilon^S E \tag{42}$$

where the variables T, c, E, S, e, D, ε^S are the stress, Young's modulus, electric field, strain, piezoelectric constant, electric displacement, and the permittivity of the piezoelectric material. The Piezoelectric Decoupled, anisotropic model was used for the actuation models. Using this modeling, the full six-by-six elasticity matrix was defined and the boundary conditions are established [21]. Figure 13 shows the simulation results of the [0/90] laminate post-cure and post actuation where you can see the change in shape direction after the MFC has been actuated.

5.3. Actuation Experiments

The bonding process done in this research was the process suggested by the Smart Material Corp. For experimental uses, a LabView chassis was used to control the system. For experimentation, the programming voltage was increased from zero by 0.02 V DC per 45 seconds until snap through was observed. An experimental snapthrough laminate is shown in Figure 14 and the experimental set-up is shown in Figure 15.

After snap through of the laminate to the orthogonal cylindrical shape II, the voltage load is removed to assure the cylindrical shape II is stable. If the shape was unstable, removing the load (voltage) would cause the laminate to snap back to cylindrical shape I. In Figure 16, the transverse actuated displacement is presented in the second picture.

After running the experimentation process a number of times and experiencing snap through, it was observed for the rectangular [0/90] laminate, an estimated voltage at snap through was 391 V DC. This voltage magnitude was then used in the numerical model to get better results for comparisons. Comparing to the analytical predictions of the post-cure and actuated shape, the experimental results are similar. The same procedure was done with the square [0/90] laminate, seen in Figure 17. The estimated voltage at snap through was 951.45 V DC, which was close to three times more than the snap through voltage for the rectangular laminate.

After cooling to room temperature and bonding the MFC actuator to the deformed laminates, the piezoelectric actuation experiments are ready to be done. The [0/90] laminates were done first for piezoelectric actuation. As seen in previous chapters for the [0/90] laminate, the snap through from cylindrical shape I should cause the laminate to have actuated displacement along the y-axis in the positive z-direction, this being cylindrical shape II. The rectangular [0/90] laminate is shown in the Figure 18.

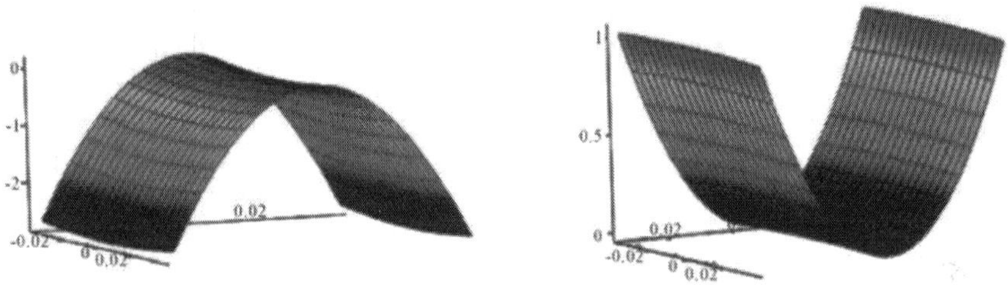

Figure 12. Predicted actuated shapes, cylindrical shape II for [0/90] and $[0_2/90_2]$.

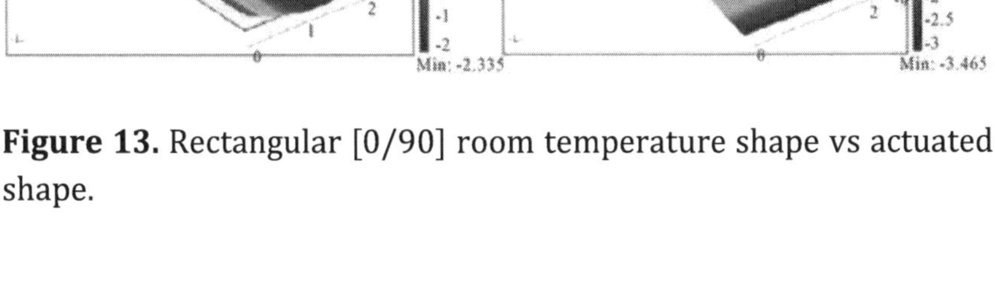

Figure 13. Rectangular [0/90] room temperature shape vs actuated shape.

As observed in earlier sections, the actuated displacement curvature is predicted by the analytical and numerical models. These predictions are then compared to the experimental results. The rectangular [0/90] laminate comparisons of the numerical and experimental results are reasonably accurate. As expected, the actuated snap through curvature of the rectangular [0/90] laminate is along the y-axis, which is the orthogonal cylindrical shape II. This proves the numerical and analytical models are reasonable prediction models for actuation. The same actuated snap through curvature is expected for the square [0/90] laminate shown in Figures 18 and 19.

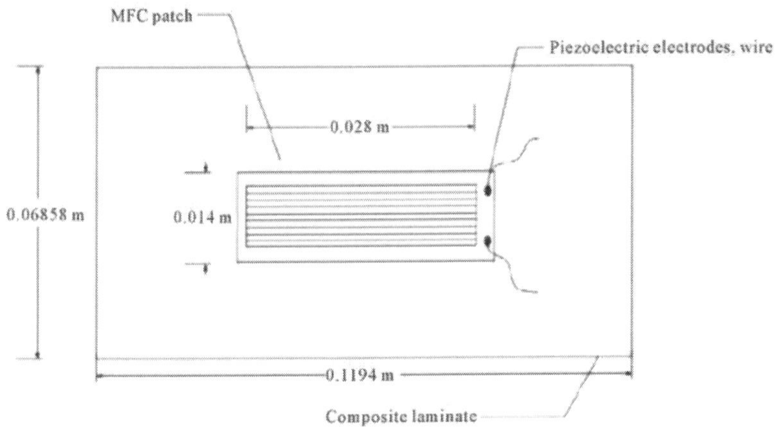

Figure 14. Experimental MFC patch on rectangular laminate.

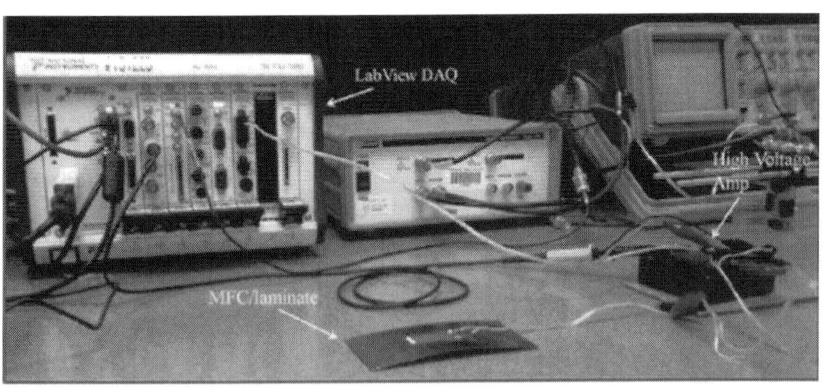

Figure 15. Experimental set-up.

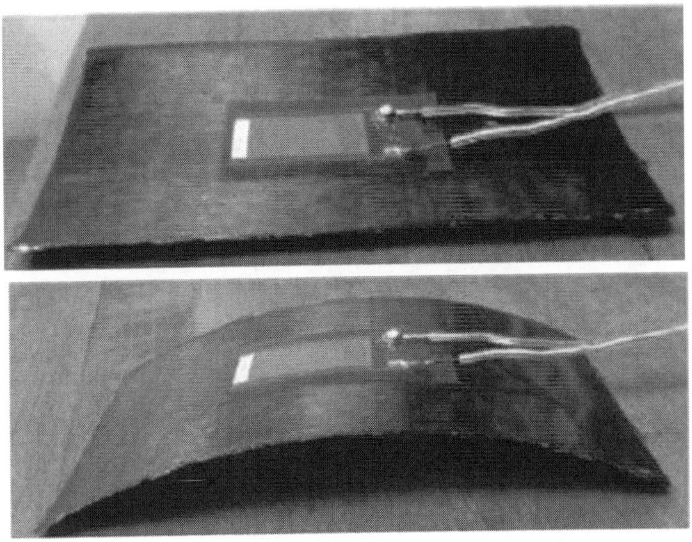

Figure 16. Cylindrical shapes I, II of the rectangular [0/90].

Due to the size and flexibility of the laminate, when using the FARO gage CMM to obtain data points the laminate will tend to move. An uncalibrated and unspecified load was required for the FARO gage to capture the deformation date of the flexible laminates. Although a laser profilometer would serve as a better tool for shape and deformation characterization, none were available. The data obtained from the experimental actuation therefore will not be as precise as the FEA ideal models. When comparing the actual experimental result to the FEA model, this error can be seen. Showing similar actuated shapes as the analytical model suggested, the square and rectangular cross-ply laminates behaved as expected. This was done by using the preliminary methods and techniques for modeling the cure process. With the results from those preliminary models, the piezoelectric actuation effect can be more accurately modeled. Modeling the cure process first for preliminary measures of the piezoelectric actuation allowed these methods and techniques to serve as reasonable predictions for the postcure room temperature shapes which is imperative to accurately modeling piezoelectric actuation of unsymmetric laminates.

6. CONCLUSION

Models were developed to characterize and predict the post-cure shape of the unsymmetric laminates for piezoelectric actuation. The initial model consisted of the CLT where assumptions of the laminate had to be made for the theory to work. Analysis were done with the CLT alone and it was observed that the laminates were all deforming to the saddle shape when cooling to room temperature. The CLT predicts that the room temperature shape of all unsymmetric laminates will be the saddle shape. Due to the failure to capture reference strains due to the large out-of-plane deformations as the laminates cool, the CLT had to be extended to account for these strains. This was done by using non-linear strain-displacement equations with the Rayleigh-Ritz technique and minimizing the total potential energy. This gave better predictions of the post-cure shape for the thin laminates. Numerical models were then done to get more accrate predictions of the post-cure room temperature shapes of the thin unsymmetric laminates. Initial pressure was added to the models to simulate the press from the mold. After the initial press was applied, the laminates cooled to room temperature. The numerical models were predicting similar room temperature shapes as the analytical models. After fabricating these laminates, it was observed that the analytical and numerical models were reasonably accurate when predicting the room temperature shapes of the thin laminates. Therefore, the models served as reasonable predictions necessary for piezoelectric actuation. Basic analytical techniques were given for modeling the piezoelectric actuation. Along with experimentation, is was shown that cure process models were able to accurately capture the steps and effects needed later in piezoelectric actuation experimentation.

Figure 17. Cylindrical shapes I, II of the square [0/90].

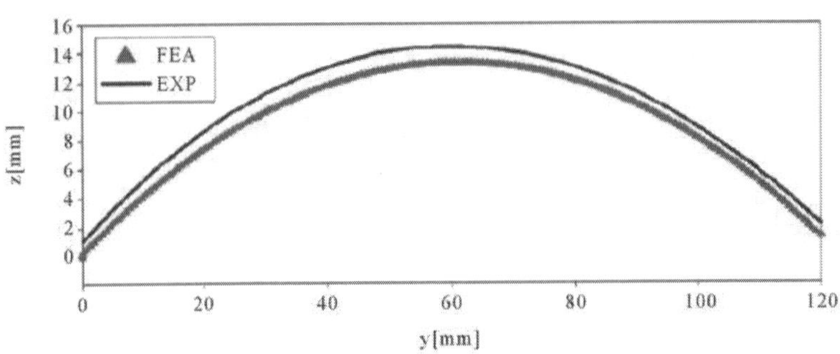

Figure 18. Rectangular [0/90] laminate actuated displacement.

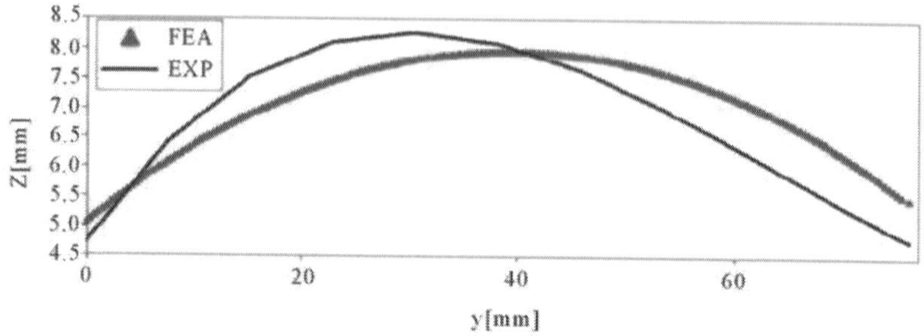

Figure 19. Square [0/90] laminate actuated displacement.

REFERENCES

1. P. Giddings, C. R. Bowen, R. Butler and H. A. Kim, "Characterisation of Actuation Properties of Piezoelectric Bi-Stable Carbon-Bre Laminates," Composites, Vol. 39, No. 4, 2008, pp. 697-703.

2. M. L. Dano and M. W. Hyer, "Snap-Through of Unsymmetric Ber-Reinforced Composite Laminates," International Journal of Solids and Structures, Vol. 39, 2001, pp. 175-198.

3. M. L. Dano and M. W. Hyer, "Thermally-Induced Deformation Behavior of Unsymmetric Laminates," International Journal of Solids and Structures, Vol. 35, No. 17, 1998, pp. 2101-2120.

4. C. R. Bowen, R. Butler, R. Jervis, H. A. Kim and A. L. T. Salo, "Morphing and Shape Control Using Unsymmetrical Composites," Intelligent Material Systems and Structures, Vol. 22, No. 18, 2007, pp. 89-98.

5. M. W. Hyer and A. Jilani, "Predicting the Deformation Characteristics of Rectangular Unsymmetric Laminated Piezoelectric Materials," Smart Material Structures, Vol. 7, No. 6, 1998, pp. 784-791.

6. M. Schlecht, K. Schulte and M. W. Hyer, "Advanced Calculation of the Room-Temperature Shapes of Thin Unsymmetric Composite Laminates," Composite Structures, Vol. 32, No. 1, 1995, pp. 627-633.

7. D. N. Bettes, I. T. Salo, C. R. Bowen and H. A. Kim, "Characterization and Modeling of the Cured Shapes of Arbitrary Layup Bi-Stable Composite Laminates," Composite Structures, Vol. 92, No. 7, 2010, pp. 1694-1700.

8. M. W. Hyer, "Stress Analysis of Ber-Reinforced Composite Materials," McGraw-Hill Companies, New York, 1998.

9. R. M. Jones, "Mechanics of Composite Materials," Taylor and Francis Group, New York, 1999.

10. M. Gigliotti, M. R. Wisnom and K. D. Potter, "Development of Curvature during the Cure of as4/8552 [0/90] Unsymmetric Composite Plates," Composites Science and Technology, Vol. 63, No. 2, 2003, pp. 187-197.

11. F. Mattioni, P. M. Weaver, K. D. Potter and M. I. Friswell, "Analysis of Thermally Induced Multi-Stable Composites," International Journal of Solids and Structures, Vol. 45, No. 2, 2008, pp. 657-675.

12. P. Portela, P. Camanho, P. Weaver and I. Bond, "Analysis of Morphing Multi Stable Structures Actuated by Piezoelectric Patches," Computers and Structures, Vol. 86, No. 3-5, 2008, pp. 347-356.

13. C. G. Diaconu, P. M. Weaver and A. F. Arrieta, "Dynamic Analysis of Bi-Stable Composite Plates," Sound and Vibration, Vol. 322, No. 4-5, 2009, pp. 987-1004.

14. K. D. Cowley and P. W. R. Beaumont, "The Measurement and Prediction of Residual Stresses in Carbon Ber/ Polymer

Composites," Composite Science and Technology, Vol. 57, No. 11, 1997, pp. 1445-1455.

15. L. Ren and A. Parvizi-Majidi, "Cured Shape of Cross-Ply Composite Thin Shells," Composite Materials, Vol. 37, No. 20, 2003, pp. 1801-1820.

16. M. Gigliotti, R. Wisnom and K. D. Potter, "Loss of Bifurcation and Multiple Shapes of Thin [0/90] Unsymmetric Composite Plates Subject to Thermal Stress," Composite Science and Technology, Vol. 64, No. 1, 2004, pp. 109- 128.

17. M. W. Hyer, "Calculations of the Room Temperature Shapes of Unsymmetric Laminates," Composite Materials, Vol. 15, 1981, pp. 296-309.

18. M. R. Wisnom, M. Gigliotti, N. Ersoy, M. Campbell and K. D. Potter, "Mechanisms Generating Residual Stresses and Distortion during Manufacture of Polymer-Matrix Composite Structures," Composites: Part A, Vol. 37, No. 4, 2006, pp. 522-529.

19. PTC, "Mathcad Users Guide," mathCAD 14.0 Edition, 2007.

20. A. J. Vizzini, "Introduction to Composite Materials," Department of Aerospace Engineering, University of Maryland, College Park, 1990.

21. COMSOL Multi-Physics 3.5a, "Structural Mechanics Module Users Guide," 2008.

22. D. J. Leo, "Engineering Analysis of Smart Material Systems," John Wiley & Sons, Inc., Hoboken, 2007.

23. M. R. Schultz, M. W. Hyer, R. B. Williams, W. K. Wilkie and D. J. Inman, "Snap-Through of Unsymmetric Laminates Using Piezocomposite Actuators," Composite Science and Technology, Vol. 66, No. 14, 2006, pp. 2442- 2448.

CHAPTER 7

Modified Differential Transform Method (DTM) Simulation of Hydromagnetic Multi-Physical Flow Phenomena from a Rotating Disk

Mohammad Mehdi Rashidi[1], Esmael Erfani[1], Osman Anwar Bég[2], Swapan Kumar Ghosh[3]

[1]Mechanical Engineering Department, Engineering Faculty of Bu-Ali Sina University, Hamedan, Iran

[2]Energy and Biomechanics Research, Aerospace Engineering, Department of Engineering and Mathematics, Sheaf Building, Sheffield Hallam University, Sheffield, UK

[3]Department of Mathematics, Narajole Raj College, West Bengal, India

ABSTRACT

A similarity solution for the steady hydromagnetic convective heat and mass transfer with slip flow from a spinning disk with viscous dissipation and Ohmic heating yields a system of non-linear, coupled, ordinary differential equations. These equations are analytically solved by applying a newly developed method namely the DTM-Padé technique which is a combination of the Differential Transform Method (DTM) and the Padé approximation. A full analytical solution is presented, as a benchmark for alternative numerical solutions. DTM-Padé is implemented without requiring linearization, discretization, or perturbation, and holds significant potential for solving strongly nonlinear differential equations which arise frequently in fluid dynamics. The regime studied is shown to be controlled by the slip parameter (g), magnetohydrodynamic body force parameter (M), Eckert (viscous heating) number (Ec), Schmidt number (Sc), Soret number (Sr), Dufour number (Du) and Prandtl number (Pr). The influence of selected parameters on the evolution of dimensionless velocity, temperature and concentration distributions is studied graphically. Increasing magnetic field (M) is found to significantly inhibit the radial (f) and tangential (g) velocities, but to accentuate the axial velocity field (h); furthermore temperature (q) and concentration (f) are both enhanced with increasing M. Increasing Soret number (Sr) acts to boost the dimensionless concentration (f). Temperatures are significantly elevated in the boundary layer regime with a rise in Eckert number (Ec). Excellent correlation between the DTM-Padé technique and numerical (shooting) solutions is achieved. The model has important applications in industrial energy systems, process mechanical engineering, electromagnetic materials processing and electro-conductive chemical transport processes.

KEYWORDS

Differential transform method, Padé approximants, Thermal-diffusion, Heat transfer, Soret effect, Boundary-layers, Hydromagnetics, Slip, Dissipation, Electromagnetic processing of materials

1. INTRODUCTION

Hydromagnetic flow with heat and mass transfer from a rotating disk is a fundamental regime of interest in diverse branches engineering including magnetic materials processing [1], industrial energy conversion [2], magnetohydrodynamic tribological systems [3,4], rotating disk voltammetry in chemical synthesis [5] etc. Static magnetic fields are frequently implemented in solidification processing for metallic alloys wherein they can be employed strategically to regulate melt flow and solidify structures in the continuous casting of steels [6]. Electromagnetic processing can significantly improve the quality of products in numerous mechanical and chemical engineering technologies. In magnetohydrodynamic disk generators, high power density can be produced with low environmental issues, and such systems are increaseingly being employed in modern thermal power systems and incorporated into enhancing industrial energy management [7,8]. Such applications have stimulated considerable interest in the mechanical engineering research community in theoretically and numerically simulating rotating disk hydromagnetic flows in the presence of various supplementary phenomena including, magnetic induction, slip, viscous heating, thermal radiation etc. Intani et al. [9] used the finite element method to study the velocity and induced magnetic fields in a rotating hydromagnetic disk generator. Aboul-Hassan and Attia [10] investigated the effects of Hall currents on hydromagnetic rotating disk flow in the presence of an axial static uniform magnetic field using a shooting method. Turkyilmazoglu [11] studied the transient hydromagnetic convection flow from a porous spinning disk in the presence of uniform normal magnetic field, using a compact, unconditionally stable, implicit spectral numerical integration algorithm.

The above studies did not consider thermo-diffusion or diffuso-thermal effects. Thermodiffusion, also called thermal diffusion or the

Soret effect corresponds to species differentiation developing in an initial homogeneous mixture subjected to a thermal gradient. On the other hand, the Soret effect has been also utilized for isotope separation and in mixture between gases with very light molecular weight, such as H_2 or He and of medium molecular weight, such as H_2 or air. In many studies the Dufour and Soret effects are neglected, on the basis that they are of a smaller order of magnitude than the effects described by Fourier's (thermal conduction) and Fick's (species diffusion) laws. However, in certain industrial processing operations, such phenomena may arise and infact exert an important role. Postelnicu [12] studied numerically the influence of a magnetic field on heat and mass transfer by natural convection from vertical surfaces in porous media considering Soret and Dufour effects. Bhargava et al. [13] used the finite element method to analyze the oscillatory reactive double-diffusive magnetohydrodynamic boundary layer flow with Soret and Dufour effects. Bég et al. [14] used shooting quadrature to simulate the influence of Soret and Dufour effects on natural convection hydromagnetic flow from a stretching sheet in porous media. Maleque and Sattar [15] have also studied MHD natural convection flow with mass transfer incorporating the influence of thermaldiffusion effects.

Dissipation effects may also contribute significantly to flow and heat/mass transfer characteristics in magnetohydrodynamic transport phenomena. These may be simulated as viscous dissipation and Joule heating (Ohmic dissipation). Attia [16] considered transpiration and Ohmic heating effects on rotating disk hydromagnetic flow with ionslip. Zueco et al. [17] used network simulation methodology to simulate the combined effects of viscous and Joule heating, thermophoresis and thermal conductivity variation on hydromagnetic convection boundary layers. Bég et al. [18] have studied viscous and Joule heating effects on transient magnetohydrodynamic plasma flow in a porous medium channel, using network simulation. Osalusi et al. [19] have employed shooting quadrature to analyze the combined Soret, Dufour, viscous and Ohmic heating effects on hydromagnetic slip flow from a spinning disk. Sibanda and Makinde [20] used a Newton-Raphson shooting method along with fourth-order Runge-Kutta integration algorithm to study the steady hydromagnetic convection from a spinning disk embedded in a porous medium with viscous and Ohmic dissipation, and additionally Hall currents. They showed that magnetic field inhibits the flow due to the opposing Lorentz force generated by the magnetic field and that both magnetic field and the Eckert number (dissipation parameter) enhance the heat transfer efficiency.

In the present study we consider a similar regime to that studied by Osalusi et al. [19], using a novel technique, the DTM-Padé method [21], which can be applied directly to nonlinear differential equations without requiring linearization, discretization and therefore, is not affected by errors associated with discretization. This technique therefore provides mechanical engineering researchers with a very robust benchmark for verifying purely numerical solutions. In this article we obtain approximate analytic solutions by the combination of the DTM and the Padé approximants for rotating hydromagnetic flow, heat and mass transfer from a spinning disk with Soret, Dufour, viscous and Joule heating effects.

Therefore, this paper has been organized as follows. In Section 2, the mathematical flow model is developed. In Sections 3 and 4 we elaborate the fundamentals of the DTM approach and the theory of Padé approximants, respectively. In Section 5, we extend the application of the DTM-Padé to construct the approximate solutions for the transformed, nonlinear, coupled, ordinary differential equations governing the rotating magnetohydrodynamic flow regime. In Section 6 the graphical results are presented with a detailed discussion. Conclusions are summarized in Section 7.

2. MATHEMATICAL MODEL

We consider the steady, hydromagnetic, incompressible Von Karman convective and slip flow due to a rotating disk in the presence of viscous dissipation and Ohmic heating with effects of thermo-diffusion (Soret effect) and diffusion-thermo (Dufour effect) effects on combined heat and mass transfer [19]. Figure 1 shows the physical model with coordinate system.

The disk rotates with constant angular velocity Ω and is placed at $z=0$, where z is the vertical axis in the cylindrical coordinates system with r and Φ as the radial and tangential axes, respectively. The components of the flow velocity are (u,v,w) in the directions of increasing (r,Φ,z), respectively, the pressure is P and the density of the fluid is ρ. T and C are the fluid temperature and concentration, respectively, and the surface of the rotating disk is maintained at a uniform temperature T_w and uniform concentration C_w. Far away from the surface, the free stream is sustained at a constant temperature T_∞, constant concentration C_∞ and at constant pressure P_∞. The fluid is assumed to be Newtonian, viscous and electricallyconducting. The external uniform magnetic field is applied perpendicular to the surface of the disk i.e. axially, and has a constant magnetic flux density B_0. Since small magnetic Reynolds number is assumed, magnetic induction effects are neglected. In addition, no electric field is assumed to exist and the Hall effect is negligible. The equations governing the motion of the hydromagnetic laminar flow of the homogeneous fluid take the following form

$$\frac{\partial u}{\partial r}+\frac{u}{r}+\frac{\partial u}{\partial z}=0 \tag{1}$$

$$u\frac{\partial u}{\partial r}-\frac{v^2}{r}+w\frac{\partial u}{\partial z}+\frac{1}{\rho}\frac{\partial P}{\partial r}$$
$$=\nu\left(\frac{\partial^2 u}{\partial r^2}+\frac{1}{r}\frac{\partial u}{\partial r}-\frac{u}{r^2}+\frac{\partial^2 u}{\partial z^2}\right)-\frac{\sigma B_0^2 u}{\rho}, \tag{2}$$

$$v\frac{\partial v}{\partial r}-\frac{uv}{r}+w\frac{\partial v}{\partial z}=\nu\left(\frac{\partial^2 u}{\partial r^2}+\frac{1}{r}\frac{\partial u}{\partial r}-\frac{v}{r^2}+\frac{\partial^2 u}{\partial z^2}\right)-\frac{\sigma B_0^2 u}{\rho}, \tag{3}$$

$$w\frac{\partial w}{\partial r}+w\frac{\partial w}{\partial z}+\frac{1}{\rho}\frac{\partial P}{\partial z}=\nu\left(\frac{\partial^2 w}{\partial r^2}+\frac{1}{r}\frac{\partial w}{\partial r}+\frac{\partial^2 w}{\partial z^2}\right), \tag{4}$$

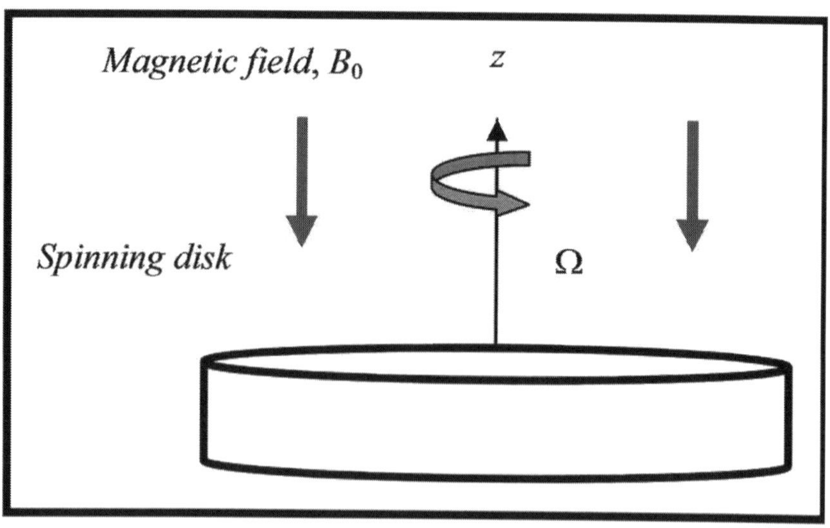

Figure 1. physical model with coordinate system.

$$u\frac{\partial T}{\partial r}+w\frac{\partial T}{\partial z}=\frac{\kappa}{\rho c_p}\left(\frac{\partial^2 T}{\partial r^2}+\frac{1}{r}\frac{\partial T}{\partial r}+\frac{\partial^2 T}{\partial z^2}\right)+\frac{\sigma B_0^2}{\rho c_p}\left(u^2+v^2\right)$$
$$+\frac{\mu}{\rho c_p}\left[\left(\frac{\partial u}{\partial z}\right)^2+\left(\frac{\partial v}{\partial z}\right)^2\right]$$
$$+\frac{Dk_T}{c_s c_p}\left(\frac{\partial^2 C}{\partial r^2}+\frac{1}{r}\frac{\partial C}{\partial r}+\frac{\partial^2 C}{\partial z^2}\right), \quad (5)$$

$$u\frac{\partial C}{\partial r}+w\frac{\partial C}{\partial z}=D\left(\frac{\partial^2 C}{\partial r^2}+\frac{1}{r}\frac{\partial C}{\partial r}+\frac{\partial^2 C}{\partial z^2}\right)$$
$$+\frac{Dk_T}{T_m}\left(\frac{\partial^2 T}{\partial r^2}+\frac{1}{r}\frac{\partial T}{\partial r}+\frac{\partial^2 T}{\partial z^2}\right), \quad (6)$$

where $\nu=\mu/\rho$ is the kinematic viscosity of the ambient fluid, κ is the thermal conductivity of heat, σ is the electrical conductivity, c_p is the specific heat at constant pressure, D is the molecular diffusion coefficient, k_T is the thermal-diffusion ratio, c_s is the concentration susceptibility and T_m is the mean fluid temperature. The last term on the right-hand side of the energy equation (5) and diffusion equation (6) signifies the Dufour or diffusion-thermo effect and the Soret or thermodiffusion effect, respectively. The boundary conditions are introduced as follows [19]:

$$u=\frac{2-\xi}{\xi}\lambda\frac{\partial u}{\partial z},\ v=r\Omega+\frac{2-\xi}{\xi}\lambda\frac{\partial v}{\partial z},\ w=0,\ T=T_w,$$
$$C=C_w \text{ at } z=0,\ u\to 0,\ v\to 0,\ T\to T_\infty,$$
$$C\to C_\infty,\ P\to P_\infty,\ \text{as } z\to\infty. \quad (7)$$

The transformed "similarity" nonlinear ordinary differential equations (9)-(13) are obtained by initially performing a non-dimensionalization [19] and subsequently by introducing a dimensionless normal distance from the disk, $\eta = Z\nu^{-1/2}$ along with the von Karman transformations:

$$U = rf(\eta),\ V = Rg(\eta),\ W = \nu_1^{\frac{1}{2}} h(\eta),$$
$$T = \theta(\eta),\ C = \varphi(\eta), \tag{8}$$

Where f, g, h, θ and ϕ are non-dimensional functions of modified dimensionless vertical (axial) coordinate η.

$$h' + 2f = 0, \tag{9}$$

$$f'' - f^2 + g^2 - f'h - Mf = 0, \tag{10}$$

$$g'' - 2fg - hg' - Mg = 0, \tag{11}$$

$$\frac{1}{\Pr}\theta'' - h\theta' + MEc\left(f^2 + g^2\right) + Ec\left(f'^2 + g'^2\right) + Du\phi'' = 0, \tag{12}$$

$$\frac{1}{Sc}\phi'' - h\phi' + Sr\theta'' = 0, \tag{13}$$

where $M = \sigma B_0^2/\rho\Omega$ is the magnetic (hydromagnetic body force) parameter, $\Pr = \rho v c_p/\kappa$ is the ordinary Prandtl number, $Ec = L^2\Omega^2R^2/c_p(T_w - T_\infty)$ is the Eckert number, $Du = D(C_w - C_\infty)k_T/c_s c_p v(T_w - T_\infty)$ is the Du-four number, $Sc = v/D$ is the Schmidt number, $Sr = D(T_w - T_\infty)k_T/v(C_w - C_\infty)$ is the Soret number and the prime symbol denotes a derivative with respect to η. The transformed boundary conditions are given as:

$$f(0) = \gamma f'(0),\ g(0) = 1 + \gamma g'(0), \tag{14}$$

$$h(0) = 0,\ \theta(0) = 1,\ \varphi(0) = 1, \tag{15}$$

$$f(\infty) = 0,\ g(\infty) = 0,\ \theta(\infty) = 0,\ \varphi(\infty) = 0, \tag{16}$$

where $\gamma = [(2-\xi)\lambda]/\xi v^{1/2}$ is the slip factor, The boundary conditions (14)-(16) imply that both the radial and tangential velocity components as well as temperature and concentration vanish sufficiently far away from the rotating disk, whereas the axial velocity component is anticipated to approach a (yet unknown) asymptotic limit for sufficiently large η-values.

3. DIFFERENTIAL TRANSFORM METHOD (DTM)

The concept of DTM was first introduced by Zhou [22] and employed to solve both linear and nonlinear initial value problems in electric circuit theory. In this method, we applied certain transformation rules. The governing differential equations and the boundary conditions of the system are transformed into a set of algebraic equations and the solution of these algebraic equations yields the desired solution of the problem. Also, DTM does not require "perturbation" parameters. The validity of DTM is independent of whether or not there exist small parameters in the considered equation. Therefore, as with other semi-analytical techniques such as the Homotopy Analysis Method [23,24] (HAM) and the Homotopy Perturbation Method (HPM) [25,26], DTM can overcome the foregoing restrictions and limitations of perturbation methods. Rashidi and Erfani [27] used DTM to solve Burgers' and nonlinear heat transfer equations demonstrating very good correlation with HAM solutions. In recent years, DTM has been successfully employed to solve many types of nonlinear problems arising in diverse branches of applied physics and engineering sciences. These include stability problems in structural dynamics [28], Lane-Emden type differential equations in astrophysical fluid dynamics and thermionic current systems [29], rotating hydromagnetic flow stability [30], water wave hydrodynamics [31], the sine-Gordon equation which arise in differential geometry and studies of propagation of magnetic flux [32], soil-structure interaction problems in earthquake engineering [33], the regularized long wave (RLW) equation in plasma wave hydromagnetics [34], boundary layer theory in aerodynamics [35], elasto-dynamics [36] and unsteady, nonlinear thermal conduction modelling in mechanical engineering components [37]. All these successful applications have extensively verified the validity, effectiveness, and flexibility of DTM.

DTM constructs for differential equations an analytical solution in the form of a power series. Furthermore, power series are not useful for large values of η, say $\eta \to \infty$. It is now well-known that the Padé approximants [38,39] have the advantage of manipulating the polynomial approximation into rational functions of polynomials. It is therefore essential to combination of the series solution, obtained by the DTM with the Padé approximant to provide an effective tool to handle boundary value problems at infinite domains. One of the first successful applications of DTM to boundary-layer flows was presented by Rashidi and Domairry [40] and Rashid [41]. Rashidi and Erfani [42] also demonstrated the applicability of the DTM approach for convection boundary layer flows over a flat plate with convective surface boundary conditions. The DTM approach has thusfar not received significant attention in rotating hydromagnetic fluid dynamics, and this is one of the objectives of the present paper. Prior to developing a DTM solution to the nonlinear rotating disk flow regime under consideration, we provide a summary of the fundamental concepts of DTM, as an aid to readers.

Consider a function $u(x)$ which is analytic in a domain T and let $x = x_0$ represent any point in the T The function $u(x)$ is then represented by a power series whose centre is located at x_0. The differential transform of the function $u(x)$ is given by:

$$U(k) = \frac{1}{k!}\left[\frac{\mathrm{d}^k u(x)}{\mathrm{d}x^k}\right]_{x=x_0}, \tag{17}$$

where $u(x)$ is the original function and $U(k)$ the transformed function. The inverse transformation is defined as follows

$$u(x)=\sum_{k=0}^{\infty}(x-x_0)^k U(k). \tag{18}$$

Combining Equations (17) and (18), we arrive at:

$$u(x)=\sum_{k=0}^{\infty}\frac{(x-x_0)^k}{k!}\left[\frac{\mathrm{d}^k u(x)}{\mathrm{d}x^k}\right]_{x=x_0}. \tag{19}$$

Considering Equation (19), it is apparent that the concept of the DTM is derived from Taylor series expansion. However, DTM does not evaluate the derivatives symbolically. In practical applications, the function $u(x)$ is expressed by a finite series and Equation (18) can be rewritten as follows

$$u(x)\cong\sum_{k=0}^{m}(x-x_0)^k U(k), \tag{20}$$

which implies that $\sum_{k=m+1}^{\infty}(x-x_0)^k U(k)$ is negligibly small. Usually, the value of m is decided by convergence of the series coefficients. We have documented operations for differential transformed functions about the point $x=0$ in Table 1 (see Appendix 1) and we assume that $x_0=0$ in the following sections.

4. THE PADÉ APPROXIMANTS

Suppose that we are given a power series $\sum_{i=0}^{\infty} a_i x^i$, representing a function $f(x)$, so that

$$f(x)=\sum_{i=0}^{\infty} a_i x^i, \tag{21}$$

a Padé approximant is a rational fraction and the notation for such a Padé approximant is [38,39]:

$$[L,M]=\frac{P_L(x)}{Q_M(x)}, \tag{22}$$

where $P_L(x)$ is a polynomial of degree at most L and $Q_M(x)$ is a polynomial of degree at most M. we have

$$f(x)=a_0+a_1x+a_2x^2+a_3x^3+a_4x^4+\cdots, \tag{23}$$

$$P_L(x) = p_0 + p_1 x + p_2 x^2 + p_3 x^3 + \cdots + p_L x^L, \tag{24}$$

$$Q_M(x) = q_0 + q_1 x + q_2 x^2 + p_3 x^3 + \cdots + q_M x^M. \tag{25}$$

Inspection of Equation (22) reveals that there are $L+1$ numerator coefficients and $M+1$ denominator coefficients. Since we can clearly multiply the numerator and denominator by a constant and leave $[L,M]$ unchanged, we impose the normalization condition:

$$Q_M(0) = 1 = q_0. \tag{26}$$

Therefore, there are $L+1$ independent numerator coefficients and M independent denominator coefficients, making $L+M+1$ unknown coefficients in all. This number suggests that normally the $[L,M]$ ought to fit the power series Equation (21) through the orders $1, x, x^2, \cdots, x^{L+M}$. Using the recommendation provided in [38,39], we know that the $[L,M]$ approximant is uniquely determined. In the notation of formal power series:

$$\sum_{i=0}^{\infty} a_i x^i = \frac{p_0 + p_1 x + p_2 x^2 + \cdots + p_L x^L}{q_0 + q_1 x + q_2 x^2 + \cdots + q_M x^M} + O\left(x^{L+M+1}\right). \tag{27}$$

By cross-multiplying Equation (27), we find that:

$$\left(a_0 + a_1 x + a_2 x^2 + a_3 x^3 + a_4 x^4 + \cdots\right)$$
$$\left(1 + q_1 x + q_2 x^2 + \cdots + q_M x^M\right)$$
$$= p_0 + p_1 x + p_2 x^2 + \cdots + p_L x^L + O\left(x^{L+M+1}\right). \tag{28}$$

These yield the following set of equations:

$$\begin{cases} a_0 = p_0, \\ a_1 + a_0 q_1 = p_1, \\ a_2 + a_1 q_1 + a_0 q_2 = p_2, \\ \vdots \\ a_L + a_{L-1} q_1 + \cdots + a_0 q_L = p_L, \end{cases} \tag{29}$$

And

$$\begin{cases} a_{L+1} + a_L q_1 + \cdots + a_{L-M+1} q_M = 0, \\ a_{L+2} + a_{L+1} q_1 + \cdots + a_{L-M+2} q_M = 0, \\ \vdots \\ a_{L+M} + a_{L+M-1} q_1 + \cdots + a_L q_M = 0, \end{cases} \tag{30}$$

where $a_n = 0$ for $n < 0$ and $q_j = 0$ for $j > M$. If Equations (29) and (30) are non-singular, then a direct solution is tangible:

$$[L,M] = \frac{\begin{vmatrix} a_{L-m+1} & a_{L-m+2} & \cdots & a_{L+1} \\ \vdots & \vdots & \ddots & \vdots \\ a_L & a_{L+1} & \cdots & a_{L+M} \\ \sum_{j=M}^{L} a_{j-M}x^j & \sum_{j=M-1}^{L} a_{j-M+1}x^j & \cdots & \sum_{j=0}^{L} a_j x^j \end{vmatrix}}{\begin{vmatrix} a_{L-m+1} & a_{L-m+2} & \cdots & a_{L+1} \\ \vdots & \vdots & \ddots & \vdots \\ a_L & a_{L+1} & \cdots & a_{L+M} \\ x^M & x^{M-1} & \cdots & 1 \end{vmatrix}}, \quad (31)$$

If the lower index on a sum exceeds the upper, the sum is replaced by zero. Alternate forms are:

$$\begin{aligned}[L,M] &= \sum_{j=0}^{L-M} a_j x^j + x^{L-M+1} w_{L,M}^T W_{L,M}^{-1} w_{L,M} \\ &= \sum_{j=0}^{L+n} a_j x^j + x^{L+n+1} w_{(L+M),M}^T W_{L,M}^{-1} w_{(L+n),M}, \end{aligned} \quad (32)$$

$$\mathrm{W}_{L,M} = \begin{bmatrix} a_{L-M+1} - xa_{L-M+2} & \cdots & a_L - xa_{L+1} \\ \vdots & \ddots & \vdots \\ a_L - xa_{L+1} & \cdots & a_{L+M+1} - xa_{L+M} \end{bmatrix}, \quad (33)$$

$$\mathrm{w}_{L,M} = \begin{bmatrix} a_{L-M+1} \\ a_{L-M+2} \\ \vdots \\ a_L \end{bmatrix}. \tag{34}$$

The construction of $[L,M]$ approximants involves only algebraic operations [38,39]. Each choice of L, degree of the numerator and M, degree of the denominator, leads to an approximant. The major difficulty in applying this technique is how to direct the choice in order to obtain the best approximant. This requires a criterion which dictates the choice of approximation, depending on the shape of the solution. A criterion which has worked well here is the choice of $[L,M]$ approximants such that $L=M$. Using the symbolic computation software MATHEMATICA, we directly employ the command "Padé Approximant" about the point $x=0$ to generate the Padé approximant of $f(x)$ in the following sections.

5. ANALYSIS WITH DTM-PADÉ SIMULATION

Taking the one-dimensional differential transform, from Table 1 to Equations (9)-(13), the following transformations are obtained:

$$(k+1)H(k+1)+2F(k)=0, \tag{35}$$

$$\begin{aligned}&(k+1)(k+2)F(k+2)-\sum_{r=0}^{k}F(r)F(k-r)\\&+\sum_{r=0}^{k}G(r)G(k-r)-\sum_{r=0}^{k}(k-r+1)H(r)F(k-r+1)\\&-MF(k)=0,\end{aligned} \tag{36}$$

$$\begin{aligned}&(k+1)(k+2)G(k+2)-2\sum_{r=0}^{k}F(r)G(k-r)\\&-\sum_{r=0}^{k}(k-r+1)H(r)G(k-r+1)-MG(k)=0,\end{aligned} \tag{37}$$

$$\begin{aligned}&\frac{1}{\Pr}(k+1)(k+2)\Theta(k+2)\\&-\sum_{r=0}^{k}(k-r+1)H(r)\Theta(k-r+1)\\&+MEc\sum_{r=0}^{k}F(r)F(k-r)+MEc\sum_{r=0}^{k}G(r)G(k-r)\\&+Ec\sum_{r=0}^{k}(r+1)(k-r+1)F(r+1)F(k-r+1)\\&+Ec\sum_{r=0}^{k}(r+1)(k-r+1)G(r+1)G(k-r+1)\\&+Du(k+1)(k+2)\Theta(k+2)=0,\end{aligned} \tag{38}$$

$$\begin{aligned}&\frac{1}{Sc}(k+1)(k+2)\Phi(k+2)\\&-\sum_{r=0}^{k}(k-r+1)H(r)\Phi(k-r+1)\\&+Sr(k+1)(k+2)\Theta(k+2)=0.\end{aligned} \tag{39}$$

where $F(k)$, $G(k)$, $H(k)$, $\Phi(k)$ and $\Theta(k)$ are the transformed functions of $f(\eta)$, $g(\eta)$, $h(\eta)$, $\varphi(\eta)$ and $\theta(\eta)$, respectively and are given by:

$$f(\eta) \cong \sum_{k=0}^{m} F(k)\eta^k, \tag{40}$$

$$g(\eta) \cong \sum_{k=0}^{m} G(k)\eta^k, \tag{41}$$

$$h(\eta) \cong \sum_{k=0}^{m} H(k)\eta^k, \tag{42}$$

$$\varphi(\eta) \cong \sum_{k=0}^{m} \Phi(k)\eta^k, \tag{43}$$

$$\theta(\eta) \cong \sum_{k=0}^{m} \Theta(k)\eta^k. \tag{44}$$

The differential transform of the boundary conditions Equations (14)-(16) are as follows:

$$F(0) = \gamma F(1),\ F(1) = a, \tag{45}$$

$$G(0)=1+\gamma G(1),\ G(1)=b, \tag{46}$$

$$H(0)=0, \tag{47}$$

$$\Theta(0)=1,\ \Theta(1)=c \tag{48}$$

$$\Theta(0)=1,\ \Theta(1)=d. \tag{49}$$

Substituting Equations (45)-(49) into Equations (35)- (39) and employing the recursive method we can calculate other values of $F(k),\ G(k),\ H(k),\ \Theta(k)$ and $\Theta(k)$. Hence, substituting all $F(k),\ G(k),\ H(k),\ \Theta(k)$ and $\Theta(k)$ into Equations (40)-(44), we have series solutions as given below:

$$f(\eta)\cong a\gamma+a\eta+\frac{1}{2}\left(aM\gamma+a^2\gamma^2-(1+b\gamma)^2\right)\eta^2+\cdots, \tag{50}$$

$$g(\eta)\cong 1+b\gamma+b\eta+\frac{1}{2}\left(M(1+b\gamma)+2a\gamma(1+b\gamma)\right)\eta^2+\cdots, \tag{51}$$

$$h(\eta) \cong -2a\gamma\eta - a\eta^2 + \cdots, \tag{52}$$

$$\varphi(\eta) \cong 1 + d\eta - \frac{ScSr\left(0.5\left(a^2+b^2\right)Ec\Pr + 0.5EcM\Pr\left(a^2\gamma^2+\left(1+b\gamma\right)^2\right)\right)}{Du\Pr ScSr - 1}\eta^2 + \cdots, \tag{53}$$

$$\theta(\eta) \cong 1 + c\eta + \frac{ScSr\left(0.5\left(a^2+b^2\right)Ec\Pr + 0.5EcM\Pr\left(a^2\gamma^2+\left(1+b\gamma\right)^2\right)\right)}{ScSr\left(Du\Pr ScSr - 1\right)}\eta^2 + \cdots. \tag{54}$$

The best procedure for enlarging the convergence domain of the truncated series solution is to employ the Padé approximant where the polynomial approximation is converted into a ratio of two polynomials. Without using the Padé approximant, the analytical solution as obtained by DTM, cannot satisfy the boundary conditions at infinity. It is therefore essential to combine the series solution obtained by the DTM, with the Padé approximant, to provide an effective tool for handling boundary value problems in infinite domains. Hence, applying the Padé approximation to Equations (50)-(54) and using the asymptotic boundary conditions in Equation (16) at $\eta \to \infty$ we can obtain a, b, c and d. The number of terms required is determined by the convergence of the numerical values to the desired accuracy, since a higher order approximation produces greater accuracy.

6. RESULTS AND DISCUSSION

The flow regime is controlled by a number of thermophysical parameters: magnetohydrodynamic body force parameter (M), Soret number (Sr), Dufour number (Du), Eckert number (Ec), Prandtl number (Pr) and Schmidt number (Sc) and slip parameter (g). The first six of these parameters arise in the boundary layer equations (9) to (13); the last parameter (g) arises in the boundary conditions (14). This parameter is linked to the slip of particles on the disk surface. This arises when the mean free path (l) of fluid particles is of the same order as the characteristic dimension of the flow domain; the classical "noslip" wall boundary condition is then inadequate and a Knudsen number is invoked to simulate slip effects [44]; a slip factor is effectively employed to simulate the wall slip. Further details of slip effects are provided in [45]. In the present simulations, we adopt a Prandtl number (Pr) corresponding to air at 20 Celsius i.e. Pr = 0.71. At this Prandtl number, the diluting chemical species generally encountered in mechanical engineering operations posses a Schmidt number (Sc) in the range from 0.1 to 10. Sc = 0.22 corresponds to Hydrogen; Sc = 0.6 corresponds to water vapor (which represents a diffusing chemical species of most common interest in air) and Sc = 0.75 is an accurate representation of oxygen diffusing in air. We further note that larger values of Sc correspond to higher molecular weight gases appropriate values for Sc are, for methanol diffusing in air, Sc = 1.0, and for ethylbenzene in air (Sc = 2.0), as indicated by Gebhart and Pera [43]. Sc measures the relative effectiveness of momentum and mass transport by diffusion. In all our computations, Sc ^{1} Pr, which physically implies that the thermal and species diffusion regions are of different extents, and therefore thermal and species boundary layer thicknesses will also be different. To verify the accuracy of the present DTMPadé computations, we have compared the DTM-Padé results with the results generated with a numerical shooting method, coupled with a Runge-Kutta integration scheme. Figures 2 and 3 show the radial velocity $f(\eta)$ distributions obtained by the DTM-Padé approach utilizing the diagonal Padé approximation [10] in comparison with the numerical solutions. Figure 2 show the radial velocity for the case of slip factor $\gamma = 0.2, 0.4, 0.6, 0.8$ and Figure 3 depicts the radial velocity for the case of magnetic interaction parameter, $M = 0.1, 0.2, 0.3, 0.4$. Excellent agreement is demonstrated in both figures 2 and 3 between DTM-Padé and the numerical results. For analytical solution, the convergence analysis was performed and in Equations (40)- (44), the value of m is selected equal to 20. The order of Padé approximation [L, M] [10] has sufficient accuracy; we further highlight that if the order of Padé approximation is increased, the accuracy of the solution increases. Figure 2 also shows that an enhancement in wall slip factor serves to elevate the radial flow i.e. increase f(h) values close to the disk surface, but conversely decrease them further from the wall. The rotating disk regime acts much like a fan system drawing fluid axially inward from the surrounding medium towards the disk surface. Since no transpiration occurs at the disk surface i.e. it is solid (suction = injection = 0), therefore the incoming fluid is turned and discharges in the radial direction along the disk. The centrifugal force associated with the spinning disk motion causes the outward radial flow. Radial flow therefore grows from the disk surface as we proceed in an axial direction, peaking and then descending to zero far from the disk surface in the free stream. Increasing slip coefficient will serve to enhance the radial flow at the disk surface (h = 0) since g i.e. the slip parameter effect will be maximized here; further from the disk surface as we progress into the boundary layer regime, transverse to the disk the wall slip effect will be gradually diminished. An increase in magnetic parameter (M) also acts to strongly retard the radial flow development. In the transformed radial momentum equation (10), the magnetic field term arises in the Lorentizian hydromagnetic drag force, -Mf'; this linear body force term is clearly accentuated with an increase in M and therefore greater magnetic field will serve to inhibit the radial flow. This is of great benefit in magnetic materials processing operations, utilizing static transverse uniform magnetic field, since it allows a strong regulation of the flow field in the vicinity of the disk. Very similar results have also been computed by Osalusi et al. [19] who used shooting quadrature and more recently by Bég et al. [45], who employed Network Simulation Methodology (NSM) and neglected Soret and Dufour effects.

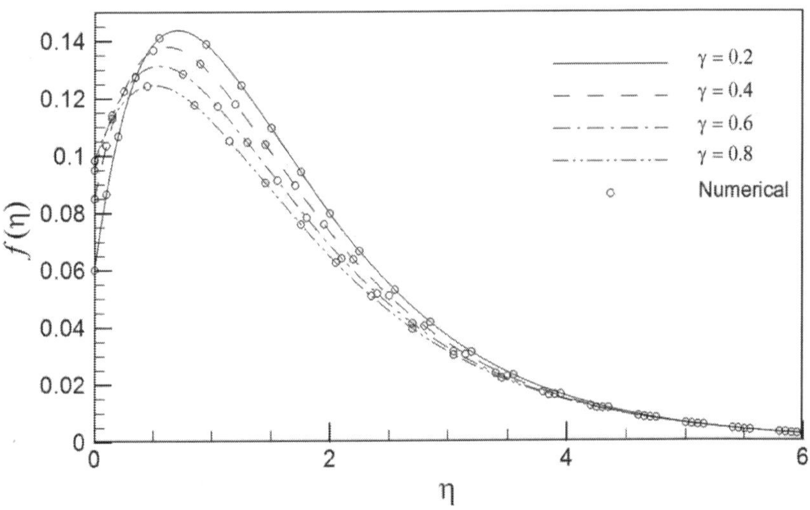

Figure 2. Analytical solutions for radial velocity, f(η) variation obtained by the DTM-Padé scheme in comparison with the numerical solutions (M = 0.2, Ec = 0.2, Du = 0.06, Sc = 0.2, Sr = 1, Pr = 0.71).

Figure 4 illustrates the effect of the Soret number (Sr) on the evolution of species (concentration) in the boundary layer regime. Sr signifies the effect of temperature gradients inducing substantial mass (species) diffusion effects. With a rise in Sr, there is a significant elevation in j values. A smooth decay in profiles is evident, from the disk surface to the free stream. Similar trends have been reported by Postelnicu [12], Bhargava et al. [13], Bég et al. [14] and Maleque and Sattar [15]. Osalusi et al. [19] also achieved similar results and we emphasize here that significant modifications in the concentration profiles are generally obtained when Du (i.e. Dufour number, which simulates the effects of concentration gradients on thermal energy flux in the regime), is non-zero. In the present case we have used Du = 0.06.

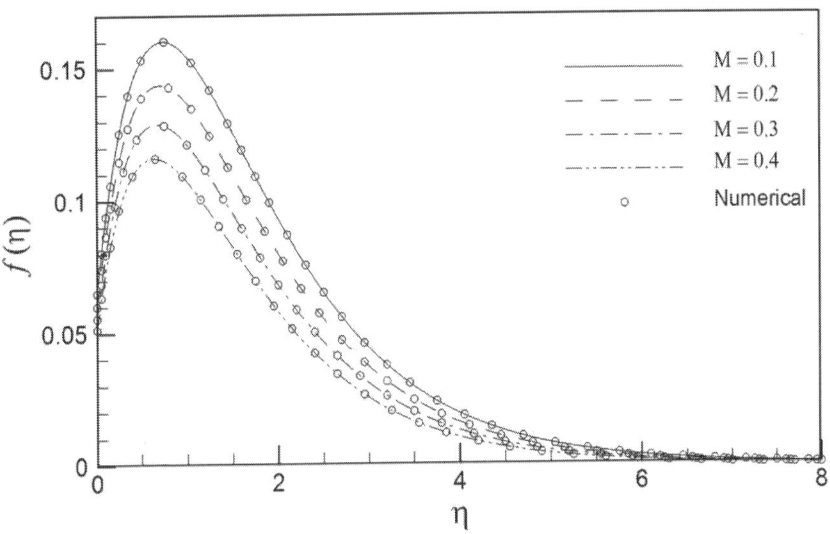

Figure 3. Analytical solutions for radial velocity, f(η) variation obtained by the DTM-Padé scheme in comparison with the numerical solutions (γ = 0.2, Ec = 0.2, Du = 0.06, Sc = 0.2, Sr = 1, Pr = 0.71).

The effects of Schmidt number (Sc) on the temperature and concentration distributions in the regime are shown in figures 5 and

6, respectively. With increasing Sc there is a slight elevation in temperature (q) magnitudes; concentration (j) magnitudes are however very strongly decreased with increasing Sc. Schmidt number embodies the relative rate of momentum diffusion to species (mass) diffusion. For Sc = 1, both momentum and species diffuse at the same rate and the momentum and species (concentration) boundary layers on the rotating disk are of the same order of thickness. For Sc < 1, as considered in our DTM-Padé computations, species diffusivity will exceed the momentum diffusivity in the regime. As such concentrations will be progressively diminished through the boundary layer. In all cases, as shown in figure 6, the profiles follow a smooth decay from the disk surface to the edge of the boundary layer, indicating that the DTM-Padé computations have converged smoothly.

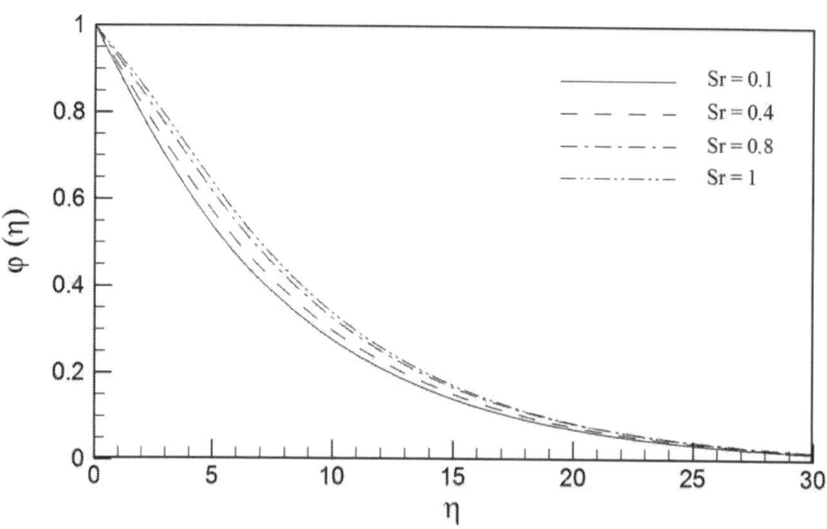

Figure 4. Influence of Sr (Soret number) on the concentration (species) profiles (γ = 0.2, M = 0.2, Du = 0.06, Ec = 0.2, Sr = 1, Pr = 0.71).

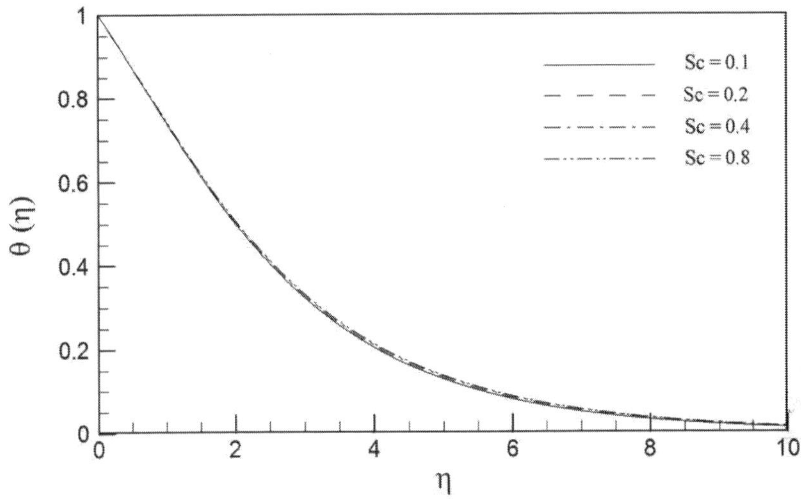

Figure 5. Influence of Sc (Schmidt number) on the temperature profiles (γ = 0.2, M = 0.2, Du = 0.06, Ec = 0.2, Sr = 1, Pr = 0.71).

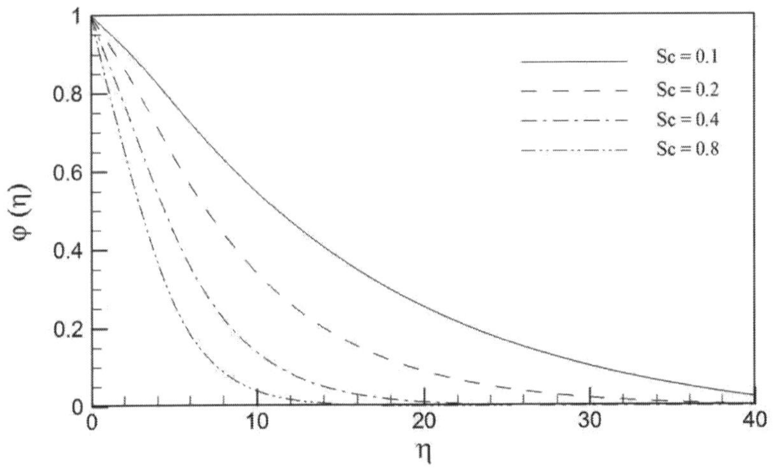

Figure 6. Influence of Sc (Schmidt number) on the concentration (species) profiles (γ = 0.2, M = 0.2, Du = 0.06, Ec = 0.2, Sr = 1, Pr = 0.71).

Figures 7 and 8 depict the response of the temperature (q) and concentration (j) functions, to various Eckert numbers (Ec). Ec represents the ratio of the kinetic energy of the flow to the boundary layer enthalpy difference as described by Schlichting [46]. Ec features in two terms in the transformed energy equation (12), viz the Joule heating (Ohmic magnetic dissipation) term, MEc ($f^2 + g^2$), and the conventional viscous heating term Ec ($f'^2 + g'^2$). Although Ec is used in high altitude rocket aero-thermodynamics (where the prescribed temperature difference is of the same order of magnitude as the absolute temperature in the free stream), in the context of low speed incompressible flows, as studied here, Ec signifies the difference between the total mechanical power input and the smaller amount of total power input which produces thermodynamically reversible effects, i.e. elevations in kinetic and potential energy. This difference constitutes the energy dissipated as thermal energy by viscous effects, i.e. work done by the viscous fluid in overcoming internal friction, hence the term viscous heating. Positive values of Ec correspond to disk cooling, i.e. loss of heat from the disk surface to the fluid; negative values imply the reverse, i.e. disk heating wherein heat is received by the disk from the fluid. In this study we restrict attention to Ec > 0. A significant increase in temperature in the boundary layer is observed (figure 7) with an increase in Ec from 0.1 through 0.2, 0.4 to 0.8. As expected there is negligible alteration in the concentration profiles (figure 8) with a change in Ec.

Figures 9-13 illustrate the distributions of radial velocity (f), tangential velocity (g), axial velocity (h), temperature (q) and concentration (j) profiles with various slip parameters, γ. In all these plots, there is a weak magnetic field acting transverse to the disk (M = 0.2). Increasing g serves to elevate the radial velocity (at or near the disk surface), but strongly decelerates the tangential flow, both at the disk surface and through the boundary layer regime. The shear-driven flow (figure 10) in the tangential (azimuthal) direction, g, is progressively retarded with increasing wall slip parameter.

Flow reversal however is never generated anywhere in the boundary layer for the radial and tangential velocity profiles i.e. values are sustained as positive. Axial velocity, h, is observed in figure 11, to be consistently negative, irrespective of the value of the slip parameter. However with increasing g the axial flow is infact accelerated i.e. values of h become more positive. The radial outflow is balanced by an axial inflow towards the rotating disk i.e. the system acts as a fan drawing fluid in. As the radial flow is suppressed, the axial flow will compensate for this and will be accelerated. Both temperature (figure 12) and concentration (figure 13) are elevated by an increase in the wall slip parameter. These results computed with DTM-Padé approach again concur with the shooting computations of Osalusi et al. [19], and the network simulation solutions by Bég et al. [45], again testifying to the accuracy of the DTM-Padé analytical procedure.

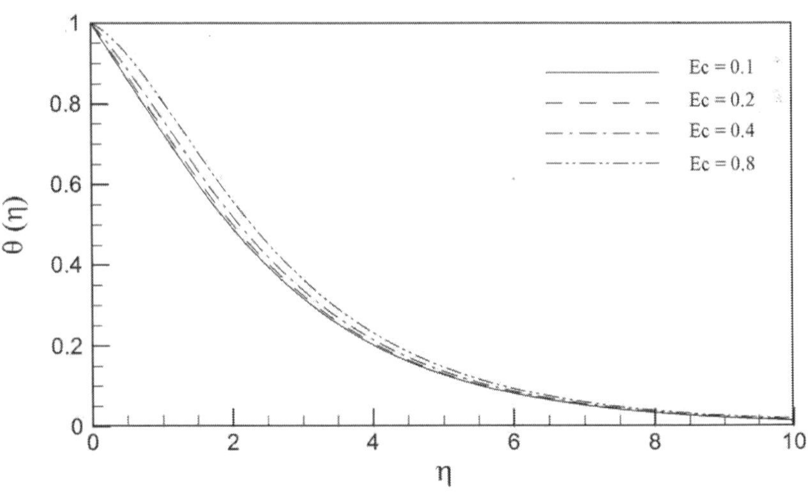

Figure 7. Influence of Ec (Eckert number) on the temperature profiles (γ = 0.2, M = 0.2, Du = 0.06, Sc = 0.2, Sr = 1, Pr = 0.71).

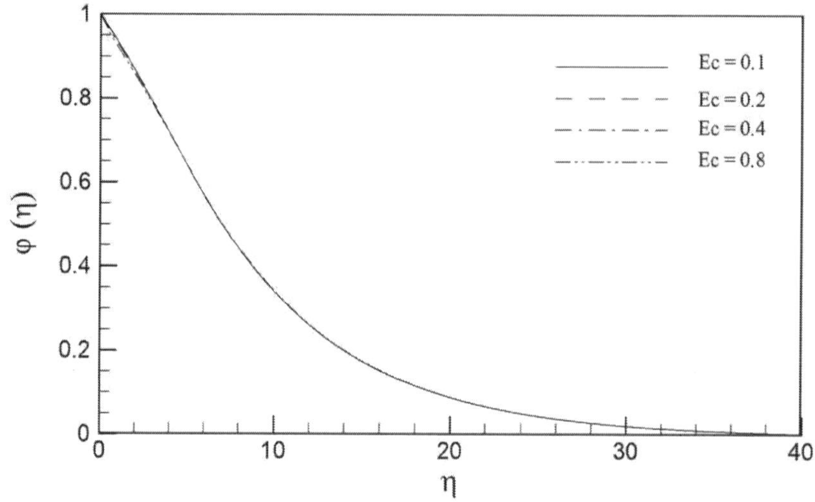

Figure 8. Influence of Ec (Eckert number) on the concentration profiles (γ = 0.2, M = 0.2, Du = 0.06, Sc = 0.2, Sr = 1, Pr = 0.71).

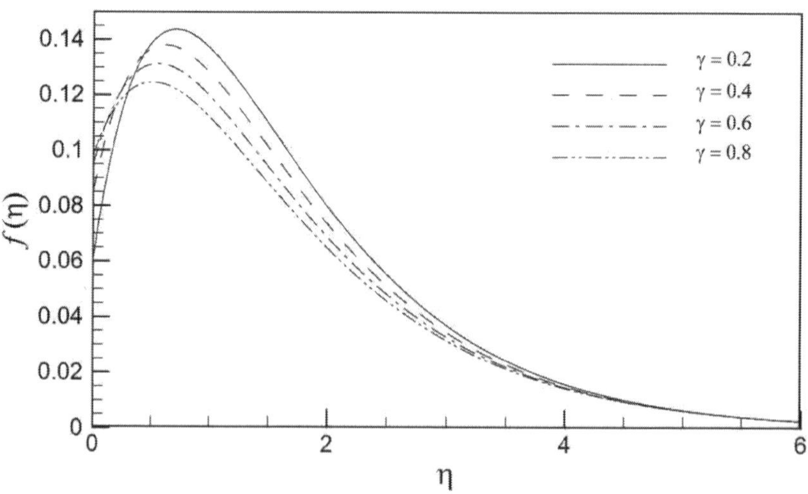

Figure 9. Influence of γ (slip parameter) on the radial velocity profiles (M = 0.2, Ec = 0.2, Du = 0.06, Sc = 0.2, Sr = 1, Pr = 0.71).

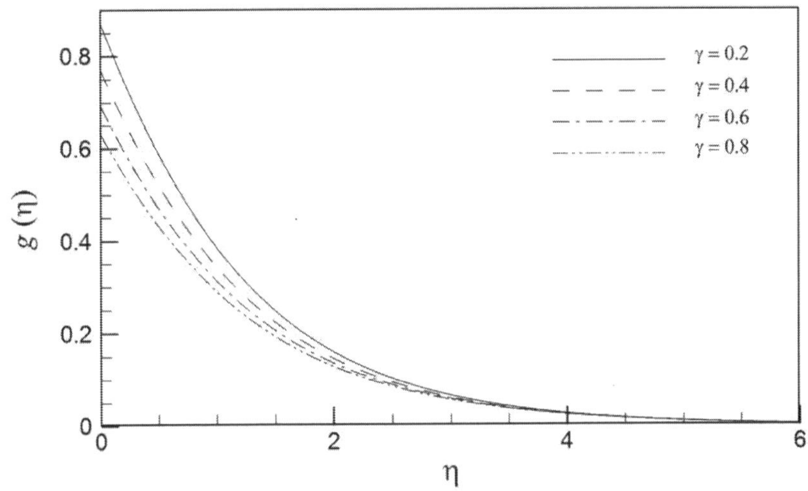

Figure 10. Influence of γ (slip parameter) on the tangential (azimuthal) velocity profiles (M = 0.2, Ec = 0.2, Du = 0.06, Sc = 0.2, Sr = 1, Pr = 0.71).

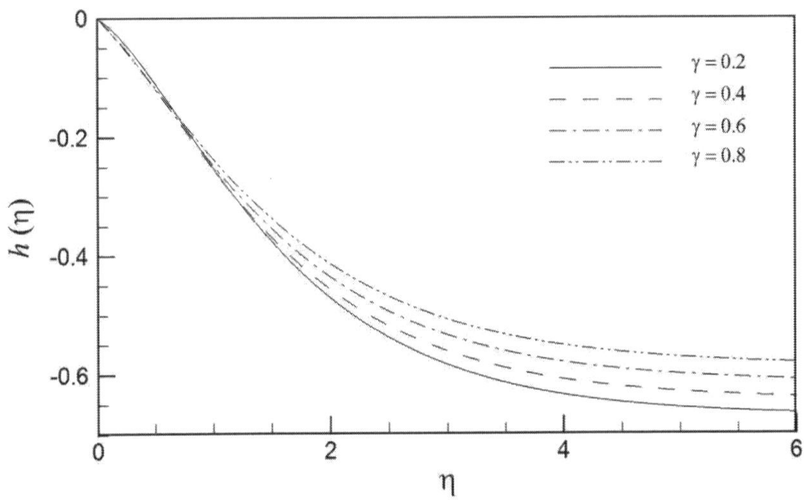

Figure 11. Influence of γ (slip parameter) on the axial velocity profiles (M = 0.2, Ec = 0.2, Du = 0.06, Sc = 0.2, Sr = 1, Pr = 0.71).

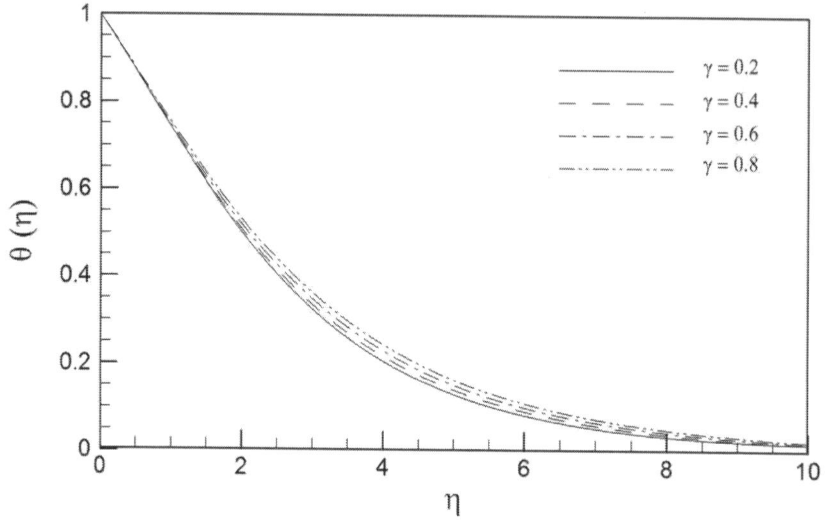

Figure 12. Influence of γ (slip parameter) on the temperature profiles (M = 0.2, Ec = 0.2, Du = 0.06, Sc = 0.2, Sr = 1, Pr = 0.71).

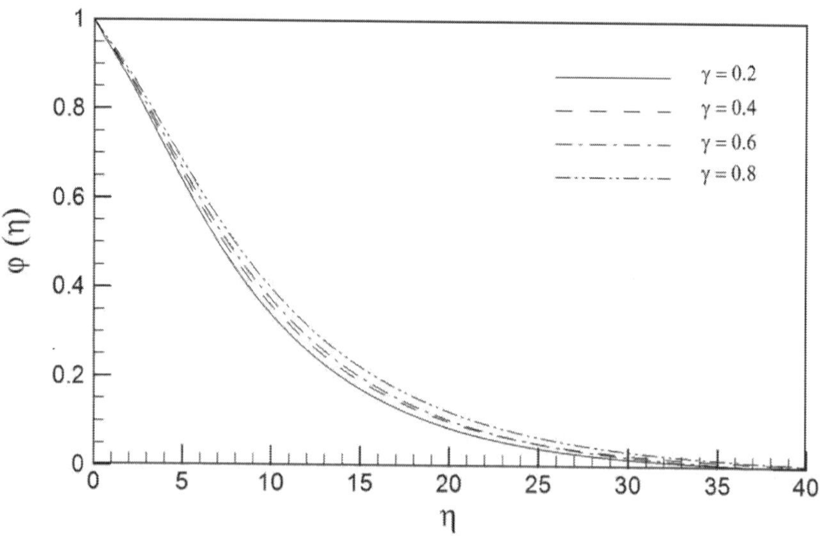

Figure 13. Influence of γ (slip parameter) on the concentration profiles, (M = 0.2, Ec = 0.2, Du = 0.06, Sc = 0.2, Sr = 1, Pr = 0.71).

Finally in figures 14-18, the effects of the magnetic parameter (M) on the radial velocity (f), tangential velocity (g), axial velocity (h), temperature (q) and concentration (j) profiles, are depicted. $M = \sigma B_0^2 / \rho\Omega$ characterizes the relative influence of the hydromagnetic drag force to the rotational (Coriolis) body force in the regime. $M > 1$ implies that hydromagnetic drag is dominant and $M < 1$ implies that rotational hydrodynamic force is dominant. For the special case of $M = 1$, both magnetohydrodynamic body force and rotational body force will have the same order of magnitude. M features in the momentum conservation equations (10) and (11) and the heat conserva tion (energy) equation (12). In equations (10) and (11), the Lorentzian magnetohydrodynamic drag terms, –Mf and –Mg, are linear. An increase in M will therefore elevate these hydromagnetic drag forces and will act to impede the radial (figure 14) and tangential (figure 15) velocities i.e. decelerate both the radial and azimuthal flow. Conversely since the applied magnetic field acts along the negative direction of the axis normal to the disk surface i.e. in the reverse h-direction (or z-direction), it will not exert a component in the axial direction (the Lorentian force acts only at right angles (i.e. transversely) to the direction of the applied magnetic field, B_0). The destruction in radial and azimuthal (tangential) momentum is compensated for by a significant boost in axial momentum. In consistency with this, we observe in figure 16 that although h values are always negative (indicating flow reversal in the axial direction), they become increasingly positive with greater M values. Hence magnetic field serves to stifle axial backflow and accelerate the flow. This trend has significant potential in magnetohydrodynamic disk energy generators and also in chemical engineering operations involving transverse magnetic fields, since the flow can be retarded effectively in the radial and tangential directions, and simultaneously enhanced in the axial direction, with the imposition of a static transverse magnetic field. In figures 17 and 18 it is evident also that temperature and concentration are strongly elevated with an increase in magnetic parameter, M. With the action of radial and azimuthal Lorentzian hydromagnetic drag forces, extra work is required to drag the fluid against the action of the magnetic field. This supplementary work is dissipated as thermal energy which causes a heating of the boundary layer regime i.e. elevation in temperatures. This effect also induces greater species diffusion in the regime and elevates the concentration in the boundary layer regime adjacent to the rotating disk. In chemical materials mixing/processing operations, therefore a stronger magnetic field will enhance concentration distributions which may be required to alter the constitution of the final product.

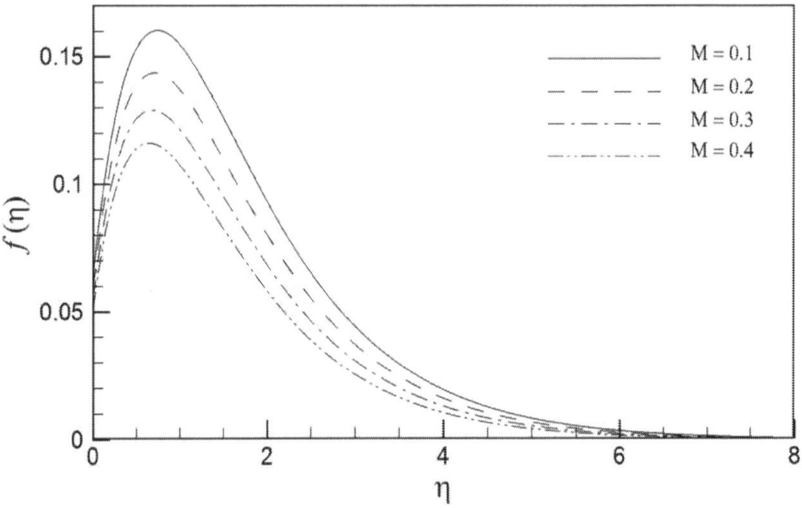

Figure 14. Influence of M (magnetic parameter) on the radial velocity profiles, (γ = 0.2, Ec = 0.2, Du = 0.06, Sc = 0.2, Sr = 1, Pr = 0.71).

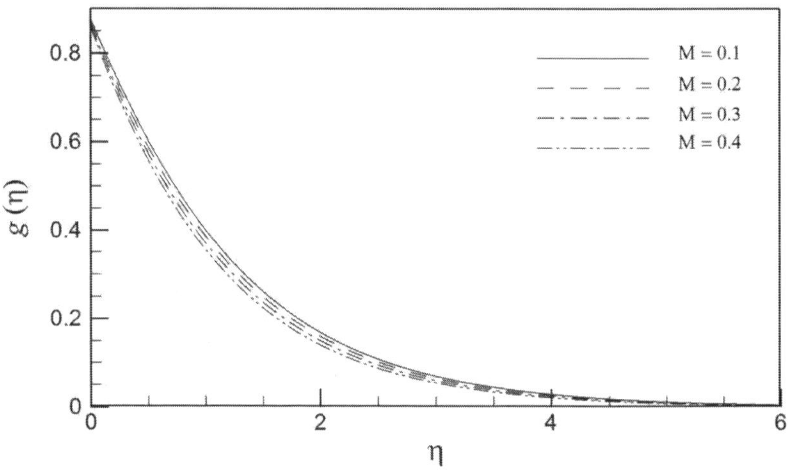

Figure 15. Influence of M (magnetic parameter) on the tangential velocity profiles, (γ = 0.2, Ec = 0.2, Du = 0.06, Sc = 0.2, Sr = 1, Pr = 0.71).

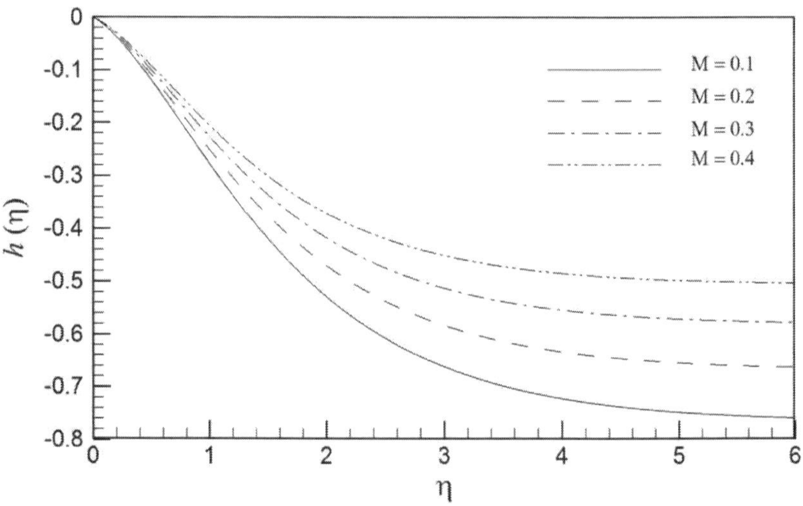

Figure 16. Influence of M (magnetic parameter) on the axial velocity profiles, (γ = 0.2, Ec = 0.2, Du = 0.06, Sc = 0.2, Sr = 1, Pr = 0.71).

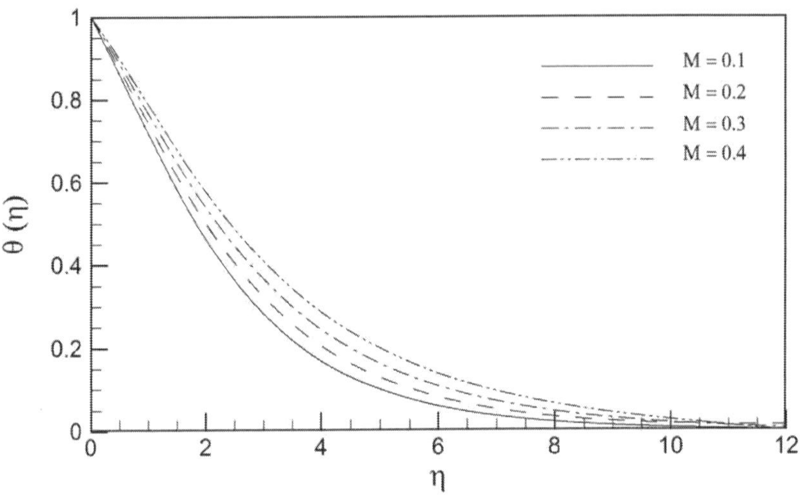

Figure 17. Influence of M (magnetic parameter) on the temperature profiles, (γ = 0.2, Ec = 0.2, Du = 0.06, Sc = 0.2, Sr = 1, Pr = 0.71).

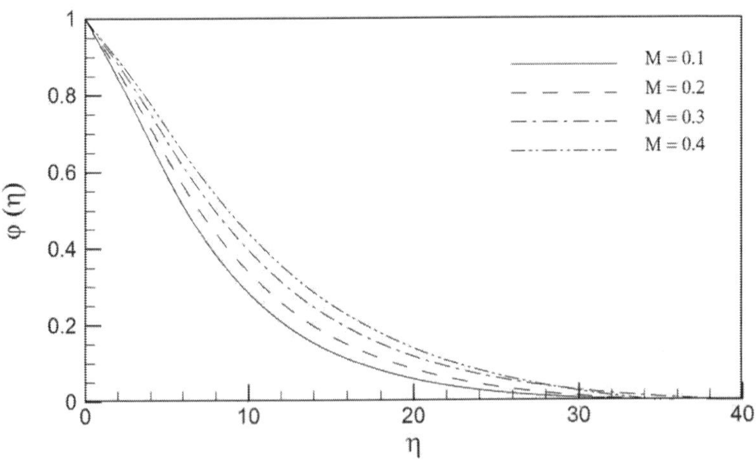

Figure 18. Influence of M (magnetic parameter) on the concentration profiles, (γ = 0.2, Ec = 0.2, Du = 0.06, Sc = 0.2, Sr = 1, Pr = 0.71).

7. CONCLUSIONS

In this paper, the DTM-Padé semi-analytical procedure has been employed to investigate combined heat and mass transfer of a steady MHD convective and slip flow due to a rotating disk with viscous dissipation, Ohmic heating, Soret and Dufour effects. The DTM technique combined with Padé approximants has been shown to be very accurate at solving the complex nonlinear differenttial equations describing the flow. DTM-Padé methodology therefore holds significant promise for nonlinear flow dynamics problems in process mechanical engineering. The method has been applied directly without requiring linearization, discretization, or perturbation. The obtained results demonstrate the reliability of the algorithm and encourage its application to other flow domains, such as non-Newtonian systems which are generally nonlinear. The results have also shown that with an increase in magnetic parameter (M), radial velocity is suppressed and tangential velocity is decreased, whereas axial velocity is elevated. Increasing Eckert number (Ec i.e. dissipation parameter) is found to elevate temperatures in the boundary layer but exerts a negligible effect on concentrations. Increasing Soret number (Sr) acts to enhance concentration distributions. With inscreasing wall slip parameter (g), radial velocity is shown to be accelerated near the disk surface and axial velocity is also elevated, whereas the tangential velocity is consistently decelerated. Furthermore there is a strong increase in temperature and concentration values in the boundary layer regime with increasing wall slip parameter. The current problem has been confined to steady flow. Future studies will examine transient rotating disk flows and also consider a variety of rheological flow and nanofluid models [47]; the results of these investigations utilizing the DTM-Padé will be communicated imminently. Extensive validation of the DTM-Padé seminumerical code developed by the authors is described also in [47].

ACKNOWLEDGEMENTS

The authors wish to express their gratitude to the reviewer for his useful comments which have helped to improve the present article.

REFERENCES

1. P. O. Tsatsin and V. P. Beskachko, "The Influence of Magnetic Field on the Inertial Deposition of a Particle on a Rotating Disk," 13th International Conference on Liquid and Amorphous Metals: Special Volume, Journal of Physics: Conference Series, Vol. 98, 2008, pp. 1-4.

2. A. Arikoglu, I. Ozkol and G. Komurgoz, "Effect of Slip on Entropy Generation in a Single Rotating Disk in MHD Flow," Applied Energy, Vol. 85, No. 12, 2008, pp. 1225- 1236.

3. E. A. Salem, M. F. Khalil and S. A. Hakim, "Thermohydrodynamic Lubrication of Rotating Magnetohydrodynamic Thrust Bearings," Wear, Vol. 62, No. 2, 1980, pp. 337-348.

4. J. Zueco and O. A. Bég, "Network Numerical Analysis of Hydromagnetic Squeeze Film Flow Dynamics between Two Parallel Rotating Disks with Induced Magnetic Field Effects," Tribology International, Vol. 43, No. 3, 2010, pp. 532-543.

5. G. Hinds, F. E. Spada, J. M. D. Coey, T. R. N. Mhíocháin and M. E. G. Lyons, "Magnetic Field Effects on Copper Electrolysis," Journal of Physical Chemistry B, Vol. 105, No. 39, 2001, pp. 9487-9502.

6. R. Bessaih, P. Marty and M. Kadja, "Numerical Study of Disk Driven Rotating MHD Flow of a Liquid Metal in a Cylindrical Enclosure," Acta Mechanica, Vol. 135, No. 3-4, 1999, pp. 153-167.

7. N. Harada, H. Yamasaki and S. Shioda, "High Enthalpy Extraction from a Helium-Driven Disk Magnetohy-drodynamic Generator," American Institute of Aeronautics and Astronautics (AIAA) Journal of Propulsion and Power, Vol. 5, No. 3, 1989, pp. 353-357.

8. A. Liberati and Y. Okuno, "Influence of Anode-Region Boundary-Layer Separation on Disk MHD-Generator Performance," IEEE Transactions on Plasma Science, Vol. 35, No. 5, 2007, pp. 1588-1597.

9. P. Intani, T. Sasaki, T. Kikuchi and N. Harada, "Analysis of Disk AC MHD Generator Performance by Finite Element Method," Journal of Plasma Fusion Research (Japan), Vol. 9, 2010, pp. 580-585.

10. A. L. Aboul-Hassan and H. A. Attia, "Flow Due to a Rotating Disk with Hall Effect," Physics Letters A, Vol. 228, No. 4-5, 1997, pp. 286-290.

11. M. Turkyilmazoglu, "Heat and Mass Transfer on the Unsteady Magnetohydrodynamic Flow Due to a Porous Rotating Disk Subject to a Uniform Outer Radial Flow," American Society of Mechanical Engineers (ASME) Journal of Heat Transfer, Vol. 132, 2010, pp. 1-6.

12. A. Postelnicu, "Influence of a Magnetic Field on Heat and Mass Transfer by Natural Convection from Vertical Surfaces in Porous Media Considering Soret and Dufour Effects," International Journal of Heat and Mass Transfer, Vol. 47, 2004, pp. 1467-1472.

13. R. Bhargava, R. Sharma and O. A. Bég, "Oscillatory Chemically-Reacting MHD Free Convection Heat and Mass Transfer in a Porous Medium with Soret and Dufour Effects: Finite Element Modeling," Inernational Journal of Applied Mathematics and Mechanics, Vol. 5, 2009, pp. 15-37.

14. O. A. Bég, A. Y. Bakier and V. R. Prasad, "Numerical Study of Free Convection Magnetohydrodynamic Heat and Mass Transfer from a Stretching Surface to a Saturated Porous Medium with Soret and Dufour Effects," Computational Materials Science, Vol. 46, No. 1, 2009, pp. 57-65.

15. K. A. Maleque and M. A. Sattar, "Similarity Solution of MHD Free-Convective and Mass Transfer Flow over an Vertical Porous Plate with Thermal Diffusion Effects," Arabian Journal of Science Engineering, Vol. 1, 2002, pp. 44-55.

16. H. A. Attia, "On the Effectivness of Ion Slip and Uniform Suction or Injection on Steady MHD Flow Due to a Rotating Disk with Heat Transfer and Ohmic Heating," Chemical Engineering Communications, Vol. 194, No. 10, 2007, pp. 1396-1407.

17. J. Zueco, O. A. Bég, H. S. Takhar and V. R. Prasad, "Thermophoretic Hydromagnetic Dissipative Heat and Mass Transfer with Lateral Mass Flux, Heat Source, Ohmic Heating and Thermal Conductivity Effects: Network Simulation Numerical Study," Applied Thermal Engineering, Vol. 29, 2009, pp. 2808-2815.

18. O. A. Bég, J. Zueco and H. S. Takhar, "Unsteady Magnetohydrodynamic Hartmann-Couette Flow and Heat Transfer in a Darcian Channel with Hall Current, Ionslip, Viscous and Joule Heating Effects: Network Numerical Solutions," Comunication Nonlinear Science Numerical Simulation, Vol. 14, No. 4, 2009, pp. 1082-1097.

19. E. Osalusi, J. Side and R. Harris, "Thermal-Diffusion and Diffusion-Thermo Effects on Combined Heat and Mass Transfer of a Steady MHD Convective and Slip Flow due to a Rotating Disk with Viscous Dissipation and Ohmic Heating," International Communications in Heat and Mass Transfer, Vol. 35, 2008, pp. 908-915.

20. P. Sibanda and O. D. Makinde, "On Steady MHD Flow and Heat Transfer Past a Rotating Disk in a Porous Medium with Ohmic Heating and Viscous Dissipation," International Journal of Numerical Methods Heat Fluid Flow, Vol. 20, No. 3, 2010, pp. 269-285.

21. M. M. Rashidi, M. Keimanesh, O. A. Bég and T. K. Hung, "Magnetohydrodynamic Biorheological Transport Phenomena in a Porous Medium: A Simulation of Magnetic Blood Flow Control and Filtration," International Journal for Numerical Methods in Biomedical Engineering, Vol. 27, 2011, pp. 805-821.

22. J. K. Zhou, "Differential Transformation and Its Applications for Electrical Circuits," Huazhong University Press, Wuhan, 1986.

23. M. M. Rashidi and S. Dinarvand, "Purely Analytic Approximate Solutions for Steady Three-Dimensional Problem of Condensation Film on Inclined Rotating Disk by Homotopy Analysis Method," Nonlinear Analysis Real World Applications, Vol. 10, No. 4, 2009, pp. 2346-2356.

24. A. Mehmood, A. Ali, H. S. Takhar, O. A. Bég, M. N. Islam and L. S. Wilson, "Unsteady Von Kármán Swirling Flow: Analytical Study Using the Homotopy Method," International Journal of Applied Mathematics and Mechanics, Vol. 6, 2010, pp. 67-84.

25. M. M. Rashidi, D. D. Ganji and S. Dinarvand, "Explicit Analytical Solutions of the Generalized Burger and Burger-Fisher Equations by Homotopy Perturbation Method," Numerical Methods for Partial Differential Equations, Vol. 25, No. 2, 2009, pp. 409-417.

26. A. Rajabi, D. D. Ganji and H. Taherian, "Application of Homotopy Perturbation Method in Nonlinear Heat Conduction and Convection Equations," Physics Letters A, Vol. 360, No. 4-5, 2007, pp. 570-573.

27. M. M. Rashidi and E. Erfani, "New Analytical Method for Solving Burgers' and Nonlinear Heat Transfer Equations and Comparison with HAM," Computer Physics Communications, Vol. 180, No. 9, 2009, pp. 1539-1544.

28. C. K. Chen and S. H. Ho, "Applications of the Differential Transform to Eigenvalue Problems," Applied Mathematics and Computation, Vol. 79, No. 2-3, 1996, pp. 173- 188.

29. V. S. Erturk, "Differential Transformation Method for Solving Differential Equations of Lane-Emden Type," Mathematical and Computational Applications, Vol. 12, 2007, pp. 135-139.

30. V. S. Erturk, "Application of Differential Transformation Method to Linear Sixth-Order Boundary Value Problems," Applied Mathematical Sciences, Vol. 1, 2007, pp. 51-58.

31. M. M. Rashidi and E. Erfani, "Travelling Wave Solutions of WBK Shallow Water Equations by Differential Transform Method," Advances in Theoretical and Applied Mechanics, Vol. 3, 2010, pp. 263-271.

32. J. Biazar and F. Mohammadi, "Application of Differential Transform Method to the Sine-Gordon Equation," International Journal of Nonlinear Science, Vol. 10, 2010, pp. 190-195.

33. S. Catal, "Solution of Free Vibration Equations of Beams on Elastic Soil by Using Differential Transform Method," Applied Mathematical Modelling, Vol. 32, No. 9, 2008, pp. 1744-1757.

34. Y. Keskin and G. Oturanc, "Numerical Solution of Regularized Long Wave Equation by Reduced Differential Transform Method," Applied Mathematical Sciences, Vol. 4, 2010, pp. 1221-1231.

35. L. T. Yu and C. K. Chen, "The Solution of the Blasius Equation by the Differential Transformation Method," Mathematical and Computer Modelling, Vol. 28, No. 1, 1998, pp. 101-111.

36. S. H. Ho and C. K. Chen, "Free Transverse Vibration of an Axially Loaded Non-Uniform Spinning Twisted Timoshenko Beam Using Differential Transform," International Journal of Mechanical Sciences, Vol. 48, 2006, pp. 1323-1331.

37. H. P. Chu and C. Y. Lo, "Application of the Hybrid Differential Transform-Finite Difference Method to Nonlinear Transient Heat Conduction Problems," Numerical Heat Transfer, Part A, Vol. 53, No. 3, 2008, pp. 295-307.

38. G. A. Baker and P. Graves-Morris, "Padé Approximants," Encyclopedia of Mathematics and Its Application 13, Parts I and II, Addison-Wesley Publishing Company, New York, 1981.

39. G. A. Baker, "Essential of Padé Approximants," Academic Press, London, 1975.

40. M. M. Rashidi and G. Domairry, "New Analytical Solution of the Three-Dimensional Navier-Stokes Equations," Modern Physics Letters B, Vol. 23, No. 26, 2009, pp. 3147-3155.

41. M. M. Rashidi, "The Modified Differential Transform Method for Solving MHD Boundary-Layer Equations," Computer Physics Communications, Vol. 180, No. 11, 2009, pp. 2210-2217.

42. M. M. Rashidi and E. Erfani, "A Novel Analytical Solution of the Thermal Boundary-Layer over a Flat Plate with a Convective Surface Boundary Condition Using DTM-Padé," International Conference on Applied Physics and Mathematics (ICAPM), Singapore, 2009.

43. B. Gebhart and L. Pera, "The Nature of Vertical Natural Convection Flow from the Combined Buoyancy Effects on

Thermal and Mass Diffusion," International Journal of Heat and Mass Transfer, Vol. 14, No. 12, 1971, pp. 2025-2040.

44. M. Gad-el-Hak, "The Fluid Mechanics of Microdevices: The Freeman Scholar Lecture," ASME—Journal of Fluids Engineering, Vol. 121, No. 1, 1999, pp. 5-33.

45. O. A. Bég, J. Zueco and L. M. Lopez-Ochoa, "Network Numerical Analysis of Optically-Thick Hydromagnetic Slip Flow from a Porous Spinning Disk with Radiation Flux, Variable Thermophysical Properties, and Surface Injection Effects," Chemical Engineering Communications, Vol. 198, No. 3, 2010, pp. 360-384.

46. H. Schlichting, "Boundary-Layer Theory," 7th Edition, MacGraw-Hill, New York, 1979.

47. M. M. Rashidi, O. Anwar Bég, M. Asadi and M. T. Rastegari, "DTM-Padé Modeling of Natural Convective Boundary Layer Flow of a Nanofluid Past a Vertical Surface," International Journal of Thermal & Environmental Engineering, Vol. 4, 2012, pp. 13-24. (in press)

CHAPTER 8

Extended modelling of dislocation transport-formulation and finite element implementation

**Michael M. W. Dogge1,2, Ron H. J. Peerlings2*
and Marc G. D. Geers2**

[1]Materials innovation institute, P.O. Box 5008, 2600 GA Delft, The Netherlands,

[2]Department of Mechanical Engineering,Eindhoven University of Technology, P.O. Box 513, 5600 MB Eindhoven, The Netherlands.

ABSTRACT

In the past decade, several higher-order crystal plasticity models have been developed to properly capture size effects, dislocation pile-up and patterning. Here we consider a formulation which accounts for the presence and behavior of both positive and negative dislocations in terms of densities. We derive an implicit finite element implementation for the continuum crystal plasticity model including dislocation transport, using a generalised continuum expression for the short-range dislocation interactions, by discretizing the two governing non-linear transport equations in time and space. The resulting non-linear algebraic equations are solved by an incremental-iterative solution scheme. We compare the resulting numerical solutions with discrete dislocation simulations. This analysis shows the capabilities of the implicit FEM framework to solve continuum dislocation transport in crystal plasticity with the added energetic dislocation interactions.

KEYWORDS

Finite element method Crystal plasticity Dislocation transport

The authors declare that they have no competing interests.

The theory regarding continuum dislocation transport was derived by MD, with input and suggestions from RP and MG to define the scope of the study. The implementation of dislocation transport in the continuum model and obtaining the simulation results were carried out by MD. The manuscript was drafted by MD. RP helped to draft the manuscript by providing input during the various stages of the drafting proces. Both RP and MG also contributed by critically revising the manuscript regarding both text and intellectual content. All authors read and approved the final manuscript.

1. BACKGROUND

Crystal plasticity models are nowadays employed in a wide range of engineering applications, and the finite element method is commonly adopted in numerical models employing them [1]. In particular, they allow one to study the effect of microstructural morphology of poly-crystalline materials. Relevant mechanisms such as anisotropy and plastic deformation along discrete slip planes, as found in crystalline materials, are captured in a natural fashion. However, other mechanisms such as size effects cannot be described by classical crystal plasticity [2, 3]. Therefore, much attention has recently been given to the incorporation of strain gradient effects in crystal plasticity models.

Physically, the plastic slip on glide planes is the result of the collective motion of dislocations [4]. On a continuum scale many higher-order phenomenological models exist which capture, rather than describe, the behaviour of a large collection of dislocations. The work of Nye [5] provided a basis for such continuum models by introducing a dislocation density tensor, representing the accumulation of dislocations when non-homogeneous deformation results in lattice incompatibility. The notion of Geometrically Necessary Dislocations (GNDs) was used to relate spatial gradients of strain to the presence of dislocations [6]. GNDs, emanating from strain gradients, can be used in a phenomenological way to e.g. elevate the strengthening of a slip system [7] or to increase the free energy of a system [8, 9], leading to additional balance equations for the plastic slip. Based on these theories, numerical models were developed including the effect of dislocations on the plastic behaviour of the material, either as strain gradients [10–12] or as additional degrees-of-freedom [13].

Here, we consider another class of higher-order crystal plasticity models, in which the presence and transport of dislocations is included more directly via dislocation densities. A crucial element in such models is the way they account for short-range dislocation interactions [14]. Such models may for instance be formulated using statistical mechanics arguments [15]. Here, however, we demonstrate that if the short-range interaction term is derived from an idealised dislocation configuration, it takes a slightly more general form, which apart from the usual gradient of GND density also contains a gradient of the total dislocation density. By upscaling the dislocation interactions in this idealized configuration we include all dislocation interactions in the continuum expression for the short-range stresses. This expression is used in the governing equations for the transport of dislocation densities, as adopted from models taking into account the evolution of dislocation densities [15–17].

The resulting initial boundary value problem contains two non-linear, coupled PDEs. In order to solve these equations numerically, they are discretised in space using the finite element method. In [16] the semi-discrete transport equations are solved using an explicit temporal discretisation. This however introduces the constraint of small time-steps in order to have stable numerical solutions.

A second objective of the present paper is therefore to formulate a fully implicit solution strategy, based on a backward Euler time integration. The obtained non-linear algebraic equations are then linearised and solved using a Newton-Raphson procedure. The evolution of the dislocation transport is thus simulated by an incremental-iterative scheme. Results obtained with this implementation are compared to discrete dislocation calculations in order to study its validity.

2. GOVERNING EQUATIONS FOR THE TRANSPORT OF DISLOCATION DENSITIES

2.1 Governing equations

In a crystal plasticity framework the plastic deformation is the result of plastic slip on distinct glide planes, and depends on the amount and orientation of such (active) slip systems. Using a small deformation assumption, for a single glide plane the total shear deformation γ can be expressed as:

$$\gamma = \gamma_e + \gamma_p. \tag{1}$$

Here, γe and γp represent the elastic and plastic deformation respectively. The stress due to this deformation is a result of the elastic deformation only. Adopting Hooke's law the following relation for the elastic deformation is obtained:

$$\gamma_e = \frac{\tau}{G}, \tag{2}$$

where τ is the shear stress acting on a glide plane and G the shear modulus. The framework is complete when an expression for the plastic slip γp is formulated. The choice for γp here is derived from the Orowan relation [4], which states that the plastic slip rate $\dot{\gamma}p$ is determined by the flux of dislocations:

$$\dot{\gamma}_p = b\Phi. \tag{3}$$

Here, b is the length of the Burgers vector and Φ is the dislocation flux. An appropriate expression is required for the dislocation transport on a glide plane accompanying the dislocation flux.

2.2 Dislocation balance equations

In a continuum framework one refrains from keeping track of all individual dislocations and their positions. Instead, we describe the dislocations in terms of densities. We first choose to describe a collection of positive and negative infinite straight edge dislocations by positive densities $\rho+(x,t)$,$\rho-(x,t)$, where x is the coordinate along the slip system and t indicates time. Uniformity is assumed in all directions perpendicular to the x-axis, so that a one-dimensional problem remains. The resulting equations are subsequently reformulated in terms of the total and GND densities $\rho(x,t)$ and $\kappa(x,t)$.

We consider the transport of a fixed number of positive and negative dislocations in a single slip system without creation and annihilation. The transport equations for the densities of positive and negative dislocations then read:

$$\frac{\partial \rho^+}{\partial t} + \frac{\partial \Phi^+}{\partial x} = 0 \tag{4}$$

$$\frac{\partial \rho^-}{\partial t} + \frac{\partial \Phi^-}{\partial x} = 0, \tag{5}$$

where $\Phi+$ and $\Phi-$ are the fluxes of positive and negative dislocations, respectively, which still depend on $\rho+$ and $\rho-$. To solve these transient PDEs we need initial and boundary conditions. The initial conditions are given by an initial density profile for both positive and negative densities, i.e. $\rho+(x,t=0)$, $\rho-(x,t=0)$. For an impenetrable barrier there can be no dislocation transport;

therefore at such a boundary there is no flux: $\Phi+=\Phi-=0$. In the case of a free surface the dislocation densities vanish at this boundary.

The fluxes in (4) and (5) can be expressed as:

$$\Phi^{\pm} = \rho^{\pm} v^{\pm} = \frac{\pm b \rho^{\pm}}{B} \left(\tau + \sigma_{\text{int}}^{\pm} \right), \tag{6}$$

where the + sign holds for positive dislocations and the – sign for negative. For the velocity $v\pm$ a linear drag law is adopted, such that the velocity depends linearly on the forces acting on the dislocations. Here, b is the length of the relevant Burgers vector, the sign depending on the sign of the dislocation, B is the drag coefficient, τ the shear stress acting on the glide plane and σint is the short-range dislocation interaction stress.

The shear stress τ includes the stress due to the net incompatibility introduced by a large number of dislocations, as well as any externally applied loading. These stresses are naturally recovered in a crystal plasticity framework by the incompatibility of plastic deformation [5, 18, 19]. As the focus here lies on the transport of dislocations, we take τ to be a constant, both in time and space. This is a valid assumption due to mechanical equilibrium in the material and for a stress-controlled deformation. The short-range interaction stress is the continuum equivalent of the interaction stresses between individual dislocations. To close the PDEs (4) and (5), we still need to establish the dependence of σshort on the dislocation densities ρ+ and ρ−.

2.3 Upscaling of the short-range interaction stress

The short-range dislocation interactions can be described by local gradients in the GND density, as formulated in [15, 16] using dislocation pair correlations. However, here we generalise the model by allowing for all dislocation gradients to contribute to the interaction stress. To formulate our continuum expression for the short-range internal stress we take into account the interactions between all dislocations as follows.

The starting point of the derivation is an idealized dislocation configuration of a single slip system in an infinite medium, containing infinite positive and negative dislocation walls with a constant vertical spacing h. The periodic dislocation wall (tilt wall) configuration studied here does not generate any long range stress, and is therefore suitable to characterise the short-range nature of the interaction stresses [14]. In this analysis we do not take into account dislocation sources and we avoid annihilation by assuming that no two dislocations of opposite sign exist on the same glide plane. However, the horizontal distribution of positive and negative walls can be arbitrary. Furthermore, we consider all walls to be mobile.

We consider the total interaction stress on both a positive and a negative wall. If there are many walls within a horizontal distance of $\pm h$ from a particular wall, i.e. the horizontal distance between walls is much smaller than the vertical spacing h within the wall, we can assume a continuous distribution of dislocation walls. Then, we can rewrite the discrete sum of individual interaction stresses in a continuum equivalent for the total interaction stresses on a positive and negative wall as follows:

$$\sigma_{\text{int}}^{\pm}(x) = \int_{-\infty}^{\infty} h\rho^{+}(x-\xi)\sigma^{\pm+}(\xi)d\xi + \int_{-\infty}^{\infty} h\rho^{-}(x-\xi)\sigma^{\pm-}(\xi)d\xi. \tag{7}$$

In this expression, $\sigma\pm+$ and $\sigma\pm-$ are the interaction stresses acting on a positive or negative dislocation wall due to another, positive or negative wall respectively, at a distance ξ to the left. The integral runs from $-\infty$ to ∞, taking into account all interactions and assuming a continuous distribution of dislocations with infinitesimally small distances. The linear densities $h\rho\pm$ are a measure for the local number of walls.

The shear stress of an infinite periodic wall of positive edge dislocations is given by [20]:

$$\sigma = \frac{\pi G b}{(1-\nu)h} \frac{\tilde{x}(\cosh 2\pi\tilde{x}\cos 2\pi\tilde{y} - 1)}{(\cosh 2\pi\tilde{x} - \cos 2\pi\tilde{y})^2}, \quad (8)$$

with G and ν the shear modulus and Poisson's ratio respectively, and $x\sim=x/h$ and $y\sim=y/h$ the normalised distances to an arbitrary dislocation within the wall. In the special case where all positive walls have the same vertical position y=0, all negative walls are shifted with respect to them by a distance h / 2 and positive and negative walls occur alternatingly with equal horizontal spacing, as shown in Fig. 1a, the obtained configuration is a Taylor lattice. In an infinite medium this configuration is stable. In this case, the positive and negative dislocations are not pairs from the same dislocation loop, but should be thought of as coming from different sources on different slip planes. Therefore, in this region dislocations do not annihilate. In the more general case considered here, visualized in Fig. 1b, where the horizontal distribution of the walls is governed by two independent densities $\rho+$ and $\rho-$, walls of different signs do not necessarily alternate and the configuration obtained is no longer necessarily a Taylor lattice. However, the vertical separation between positive and negative dislocations is preserved. It avoids annihilation between positive and negative dislocations and the underlying assumption is that two dislocations of opposite sign on the same glide plane would annihilate so quickly, that they contribute little to plastic slip. Furthermore, dislocations on the same glide plane are likely to have been emitted by the same source. In other words: if a positive dislocation is found on a certain glide plane, adjacent dislocations on the same glide plane are more likely to be also positive than negative. It is worth pointing out that the correlation between dislocations—which admittedly may be partially exaggerated—also implies that the dislocation configuration surrounding an arbitrary positive dislocation is different from that of a negative dislocation and different interaction forces may thus be experienced by them.

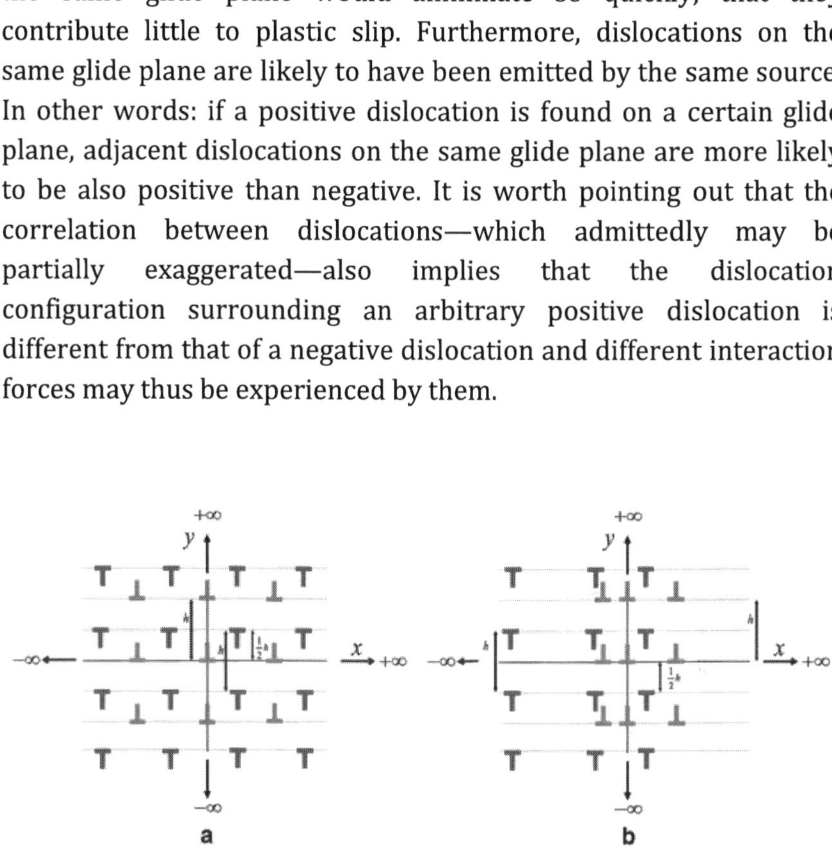

Figure 1 Taylor lattice (**A**) and a general configuration (**B**) with infinite positive and negative dislocation walls. The negative walls are shifted h / 2 with respect to the positive walls

Using this configuration, we can derive from Eq. (8):

$$\sigma^{++} = -\sigma^{--} = \sigma(\tilde{x}, \tilde{y} = 0) = \frac{\pi \bar{G} b}{h} \frac{\tilde{x}}{\sinh^2 \pi \tilde{x}} \quad (9)$$

$$\sigma^{+-} = -\sigma^{-+} = \sigma(\tilde{x}, \tilde{y} = 1/2) = \frac{\pi \bar{G} b}{h} \frac{\tilde{x}}{\cosh^2 \pi \tilde{x}}, \tag{10}$$

where $\bar{G}=G/(2(1-\nu))$.

Furthermore, the dislocation density at position $(x-\xi)$is estimated by a first order Taylor approximation around point x for both $\rho+$ and $\rho-$:

$$\rho^{\pm}(x-\xi) = \rho^{\pm}(x) - \xi \frac{\partial \rho^{\pm}}{\partial x}. \tag{11}$$

If we substitute the expressions (9), (10) and (11) in Eq. (7) and compute the resulting integrals, we obtain expressions for the total interaction stresses acting on a positive and a negative dislocation which may be written as:

$$\sigma_{\text{int}}^{\pm}(x) = \mp \frac{\bar{G} b h^2}{6} \left(2 \frac{\partial \rho^{\pm}}{\partial x} + \frac{\partial \rho^{\mp}}{\partial x} \right). \tag{12}$$

Finally, if we substitute (12) in (6) we obtain the following expression for the dislocation fluxes:

$$\Phi^{\pm} = \rho^{\pm} \frac{\pm b \tau}{B} - \rho^{\pm} \frac{\bar{G} b^2 h^2}{6B} \left(2 \frac{\partial \rho^{\pm}}{\partial x} + \frac{\partial \rho^{\mp}}{\partial x} \right). \tag{13}$$

Using (13) in (4) and (5) respectively, these PDEs are fully expressed in terms of $\rho+(x,t)$ and $\rho-(x,t)$ and we can thus calculate the evolution of the positive and negative dislocation densities.

It is worth noting that the governing equations are objective with respect to a change of sign of dislocations—as is to be expected. This can easily be verified by substituting (13) in (4) and (5), subsequently exchanging $\rho+$and $\rho-$ while replacing b by $-b$, and observing that the same governing equations are recovered.

2.4 Formulation in terms of total dislocation and GND densities

In order to more easily relate the dislocation densities to relevant and meaningful (macroscopic) properties, e.g. incompatibility and hardening, we rewrite the dislocation densities in terms of the total dislocation density ρ, and the Geometrically Necessary Dislocation density κ. The GNDs are dislocations which are required to conform to the overall deformation, i.e. they characterize the incompatibility introduced by plastic slip [5]. The remaining dislocations are called Statistically Stored Dislocations (SSDs); they do not have a net geometrical contribution. The total density then reads:

$$\rho = \rho_{\mathrm{SSD}} + |\kappa|. \tag{14}$$

Because of this relation, we only need two quantities to describe the evolution of all dislocations. A convenient choice is the combination of ρ and κ, because they have a clearer physical meaning and they can be easily related to the densities of positive and negative dislocations:

$$\rho = \rho^{+} + \rho^{-} \tag{15}$$

$$\kappa = \rho^{+} - \rho^{-}. \tag{16}$$

Using the above relations we can determine the transport equations for ρ and κ by adding (5) to (4) and subtracting (5) from (4) respectively:

$$\frac{\partial \rho}{\partial t} + \frac{\partial \Phi_{\kappa}}{\partial x} = 0 \tag{17}$$

$$\frac{\partial \kappa}{\partial t} + \frac{\partial \Phi_{\rho}}{\partial x} = 0, \tag{18}$$

where $\Phi\kappa=\Phi++\Phi-$ and $\Phi\rho=\Phi+-\Phi-$. The latter characterises the total flow of Burgers vector, irrespective of the sign of the dislocations which carry it, and thus the plastic slip rate. The former gives the net flow of Burgers vector. Using (15) and (16) and the fluxes for positive and negative dislocations (13), the fluxes $\Phi\rho$ and $\Phi\rho$ can be expressed in terms of ρ and κ as:

$$\Phi_{\kappa} = \frac{b\tau}{B}\kappa - \frac{\bar{G}b^{2}h^{2}}{12B}\left(3\rho\frac{\partial \rho}{\partial x} + \kappa\frac{\partial \kappa}{\partial x}\right) \tag{19}$$

$$\Phi_\rho = \frac{b\tau}{B}\rho - \frac{\bar{G}b^2h^2}{12B}\left(\rho\frac{\partial\kappa}{\partial x} + 3\kappa\frac{\partial\rho}{\partial x}\right). \tag{20}$$

Eqs. (17)–(20) are a closed set of equations in terms of ρ and κ.

Based on this model we can derive other models found in the literature, by using a different assumption for the upscaling of the interaction stress. If we only take into account the interactions between walls of the same sign, i.e. σ++and σ−−, the second term in the resulting interaction stresses in (12) vanishes. This leads to an expression for the short-range interaction stress in (19) and (20) only in terms of gradients of GNDs, as in References [8, 13]. The length scale introduced in these references then equals the slip plane spacing h. Furthermore, if we limit ourselves to only interactions between nearest neighbouring dislocations instead of taking into account all interactions in x-direction, and to dislocations on a single slip plane instead of walls, we obtain the interactions of References [15, 16]. In this formulation the length-scale is no longer a constant, but depends on the local dislocation density.

3. IMPLEMENTATION

In the numerical implementation of our model we need to solve the transport equations (17) and (18) for the total dislocation and GND densities. Although these equations are of a convection-diffusion nature, no stabilization is applied here. Most stabilization techniques use the addition of artificial diffusion. Here, the expression for the short-range dislocation interactions is of a diffusive nature. If one is to study these interactions in detail, addition of stabilizing diffusion can be of influence on the results. Therefore, no stabilization is introduced and care was taken choosing a sufficiently small element size and time step to ensure stability.

3.1 Time discretisation

First, the transient terms in the transport equations are discretised using a finite difference approach. For the temporal discretisation the implicit, backward Euler method is used:

$$\frac{\rho - \rho^{(t)}}{\Delta t} + \frac{\partial \Phi_\kappa}{\partial x} = 0 \tag{21}$$

$$\frac{\kappa - \kappa^{(t)}}{\Delta t} + \frac{\partial \Phi_\rho}{\partial x} = 0. \tag{22}$$

The symbols with superscript (t) represent the values at the previous time step. For brevity we dropped the superscript (t+Δt) for the values at the end of the current time step. This implicit time integration is used for additional stability and accuracy.

3.2 Spatial discretisation

For the spatial discretisation, the Galerkin method is adopted. By multiplying Eqs. (21) and (22) by test functions $\psi\rho$ and $\psi\kappa$, and integrating the resulting expression over the domain $(0, L)$ the weighted residuals formulation is obtained. Subsequently, integration by parts is applied, resulting in the weak forms:

$$\int_0^L \left(\psi_\rho \frac{\rho - \rho^{(t)}}{\Delta t} - \frac{d\psi_\rho}{dx} \Phi_\kappa \right) dx + \psi_\rho \Phi_\kappa \Big|_0^L = 0 \tag{23}$$

$$\int_0^L \left(\psi_\kappa \frac{\kappa - \kappa^{(t)}}{\Delta t} - \frac{d\psi_\kappa}{dx} \Phi_\rho \right) dx + \psi_\kappa \Phi_\rho \Big|_0^L = 0. \tag{24}$$

The degrees-of-freedom ρ,κ and the test functions $\psi\rho,\psi\kappa$ are discretised using the same shape functions $N\sim(x)$:

$$\rho = \underset{\sim}{N}^T \underset{\sim}{\rho}, \quad \psi_\rho = \underset{\sim\rho}{\psi}^T \underset{\sim}{N}$$

$$\kappa = \underset{\sim}{N}^T \underset{\sim}{\kappa}, \quad \psi_\kappa = \underset{\sim\kappa}{\psi}^T \underset{\sim}{N}.$$

Substituting these discretised variables in the weak forms (23) and (24) and requiring that they hold for all $\psi\sim\rho$ and $\psi\sim\kappa$ we obtain:

$$\underset{\sim\rho}{R} = \int_0^L \left(\underset{\sim}{N} \frac{\rho - \rho^{(t)}}{\Delta t} - \frac{d\underset{\sim}{N}}{dx} \Phi_\kappa \right) dx + \underset{\sim}{N} \Phi_\kappa \Big|_0^L = \underset{\sim}{0} \tag{25}$$

$$\underset{\sim\kappa}{R} = \int_0^L \left(\underset{\sim}{N} \frac{\kappa - \kappa^{(t)}}{\Delta t} - \frac{d\underset{\sim}{N}}{dx} \Phi_\rho \right) dx + \underset{\sim}{N} \Phi_\rho \Big|_0^L = \underset{\sim}{0}. \tag{26}$$

Note that the dislocation densities ρ,κ and the fluxes $\Phi\rho,\kappa$ can be expressed in terms of the discretised values $\rho\sim,\kappa\sim$ by using (25), (19) and (20).

3.3 Linearisation

The expressions (19) and (20) for the fluxes are non-linear; therefore an iterative solution strategy is used to solve equations (25) and (26). Here we use the Newton-Raphson method and we thus need to linearize the equations. This implies that we substitute for ρ and κ:

$$\rho = \rho^{(i)} + \delta\rho$$
$$\kappa = \kappa^{(i)} + \delta\kappa,$$

where (i) and δ denote the estimate obtained from the previous iteration and a variation (or iterative correction), respectively. The iterative updates for the fluxes $\delta\Phi\rho$ and $\delta\Phi\kappa$ are obtained by linearising (19) and (20) with respect to the degrees-of-freedom ρ,κ, giving:

$$\delta\Phi_\kappa = \frac{b\tau}{B}\delta\kappa - \frac{\bar{G}b^2h^2}{12B}\left(3\frac{\partial\rho^{(i)}}{\partial x}\delta\rho + 3\rho^{(i)}\frac{\partial\delta\rho}{\partial x} + \frac{\partial\kappa^{(i)}}{\partial x}\delta\kappa + \kappa^{(i)}\frac{\partial\delta\kappa}{\partial x}\right)$$

$$\delta\Phi_\rho = \frac{b\tau}{B}\delta\rho - \frac{\bar{G}b^2h^2}{12B}\left(\frac{\partial\kappa^{(i)}}{\partial x}\delta\rho + 3\kappa^{(i)}\frac{\partial\delta\rho}{\partial x} + 3\frac{\partial\rho^{(i)}}{\partial x}\delta\kappa + \rho^{(i)}\frac{\partial\delta\kappa}{\partial x}\right).$$

After discretisation and linearisation we obtain the following linear system of equations for the iterative corrections:

$$\begin{bmatrix} \underset{\sim}{K}_{\rho\rho} & \underset{\sim}{K}_{\rho\kappa} \\ \underset{\sim}{K}_{\kappa\rho} & \underset{\sim}{K}_{\kappa\kappa} \end{bmatrix} \begin{bmatrix} \delta\underset{\sim}{\rho} \\ \delta\underset{\sim}{\kappa} \end{bmatrix} = -\begin{bmatrix} \underset{\sim}{R}^i_\rho \\ \underset{\sim}{R}^i_\kappa \end{bmatrix}, \tag{27}$$

with:

$$\underset{\sim}{K}_{\rho\rho} = \int_0^L \left(\frac{1}{\Delta t}\underset{\sim}{N}\underset{\sim}{N}^T + \frac{d\underset{\sim}{N}}{dx}\frac{\bar{G}b^2h^2}{4B}\frac{\partial\rho^{(i)}}{\partial x}\underset{\sim}{N}^T + \frac{d\underset{\sim}{N}}{dx}\frac{\bar{G}b^2h^2}{4B}\rho^{(i)}\frac{d\underset{\sim}{N}^T}{dx}\right)dx$$

$$\underset{\sim}{K}_{\rho\kappa} = \int_0^L \left(-\frac{d\underset{\sim}{N}}{dx}\frac{b\tau}{B}\underset{\sim}{N}^T dx + \frac{d\underset{\sim}{N}}{dx}\frac{\bar{G}b^2h^2}{12B}\frac{\partial\kappa^{(i)}}{\partial x}\underset{\sim}{N}^T + \frac{d\underset{\sim}{N}}{dx}\frac{\bar{G}b^2h^2}{12B}\kappa^{(i)}\frac{d\underset{\sim}{N}^T}{dx}\right)dx$$

$$\underset{\sim}{K}_{\kappa\rho} = \int_0^L \left(-\frac{d\underset{\sim}{N}}{dx}\frac{b\tau}{B}\underset{\sim}{N}^T + \frac{d\underset{\sim}{N}}{dx}\frac{\bar{G}b^2h^2}{12B}\frac{\partial\kappa^{(i)}}{\partial x}\underset{\sim}{N}^T + \frac{d\underset{\sim}{N}}{dx}\frac{\bar{G}b^2h^2}{4B}\kappa^{(i)}\frac{d\underset{\sim}{N}^T}{dx}\right)dx$$

$$\underset{\sim}{K}_{\kappa\kappa} = \int_0^L \left(\frac{1}{\Delta t}\underset{\sim}{N}\underset{\sim}{N}^T + \frac{d\underset{\sim}{N}}{dx}\frac{\bar{G}b^2h^2}{4B}\frac{\partial\rho^{(i)}}{\partial x}\underset{\sim}{N}^T + \frac{d\underset{\sim}{N}}{dx}\frac{\bar{G}b^2h^2}{12B}\rho^{(i)}\frac{d\underset{\sim}{N}^T}{dx}\right)dx$$

When an appropriate numerical integration scheme, e.g. Gaussian integration, to approximate the above integrals is chosen, the residuals R and the tangents K can be constructed element-wise and assembled in order to solve Eq. (27). In this work, linear shape functions are chosen, and a two-point Gauss integration scheme.

4. SIMULATIONS AND RESULTS

In this section we compare results obtained with the continuum dislocation transport equations with fully discrete simulations. The problem investigated is shown in Fig. 2, including the domain boundaries. We consider a domain of length L within an infinite elastic medium with parallel slip planes on which dislocations are organised in vertical walls with a constant vertical spacing of h. The positive and negative dislocation walls are arranged such that the negative walls are shifted with $h / 2$ with respect to the positive walls. Both the positive and negative dislocation walls can be arbitrarily distributed in the horizontal direction, as determined by their respective densities.

An external shear stress τ is applied to the system. This shear stress is considered to be constant along the slip system for both the continuum and the discrete framework, because of equilibrium and because the long-range stresses for infinite dislocation walls vanish. We can therefore focus on the transport of dislocations in single slip.

After rescaling the material parameters in the governing equations, the relevant parameters are the ratios b / h, L / h and $\tau/\bar{G}$. In the simulations a fixed ratio of b/h=1/200 is used. The ratios L / h and $\tau/\bar{G}$ vary between the simulations. The number of finite elements used is 200. For the choice of the time step size, first a characteristic time can be estimated using the dislocation velocity. This velocity is determined by the linear drag law, only using the applied stress.

Using this relation an upper bound for the time step can be estimated, i.e. $\Delta t \approx hv{-}1 \approx B/(b\tau)$. An appropriate time step size should be much lower than this value to obtain accurate results.

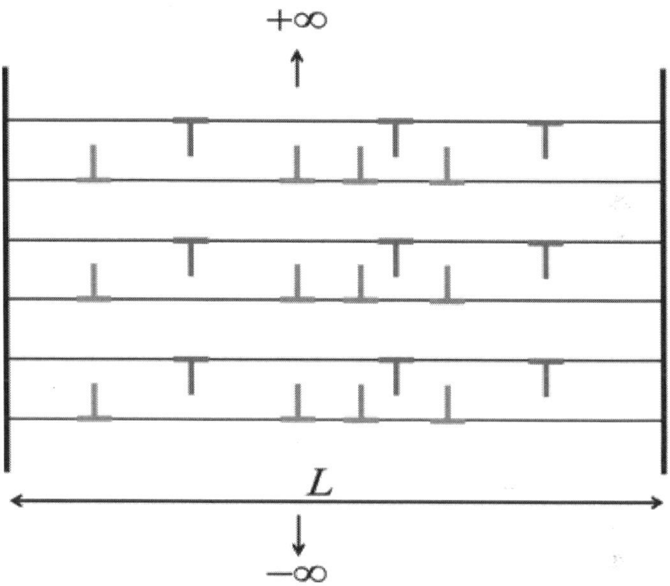

Figure 2 Infinite vertical dislocation walls with both positive and negative signs in a domain with length L. The distribution of walls is determined by the density profile. The boundaries are either impenetrable or free. An external shear stress is applied

In the discrete simulations the position of each individual wall is updated using the same linear drag law adopted in the continuum framework. The driving force in the expression for the dislocation velocity is a function of the applied shear stress on the glide plane and the interaction stress acting on the wall as a result of all other dislocations. The total interaction stress is the sum of the individual stress fields expressed in (8), with the appropriate Burgers vector sign and distance to the considered wall substituted for each dislocation wall.

Unless otherwise indicated, the boundaries of our domain are impenetrable barriers to dislocation motion. It is natural to require that the dislocation velocity is zero at these boundaries. For the continuum model this means that the boundary terms vanish, i.e. $\Phi\rho(x=0)=\Phi\rho(x=L)=\Phi\kappa(x=0)=\Phi\kappa(x=L)=0$. In the discrete dislocation simulations, immobile dislocations of the relevant sign are placed on all slip planes at such boundaries. The singular stress field emitted by them prevents the mobile dislocations within the domain from reaching the boundary.

4.1 Double pile-up of positive and negative dislocations

First, the initial dislocation configuration is chosen such that a Taylor lattice configuration is obtained as in Fig. 1a, in which positive and negative walls are evenly distributed and alternating. This corresponds in the continuum model with $\rho=\rho SSD,\kappa=0$ as the initial condition. We first apply a constant shear stress on the domain, resulting in a double pile-up against the impenetrable barriers at the two ends of the domain. The stress is chosen such that it is high enough for a negative wall to pass a positive wall. A ratio of $\tau/\bar{G}=1/20$ is adopted. The length of the domain is $L=20h$. There are 128 dislocation walls present in the domain, and the uniform initial dislocation density profile is chosen accordingly.

The evolution of the double pile-up is shown in Fig. 3. The density of SSDs, ρSSD , decreases as the positive and negative walls are moving in opposite directions due to the applied shear stress. At the same time the density of GNDs increases at the boundaries, such that a linear profile is obtained, as explored in more detail below. In order to study mesh convergence, the number of elements used in this simulation is varied. As shown in Fig. 4, when the number of elements is increased the solution converges to a certain density profile and the oscillations which are visible in the solutions for coarser discretisations disappear. Note that the number of elements giving a stable solution would be lower if a stabilization had been applied. However, as can be seen in Fig. 4, stabilization is not required to yield meaningful results. Even with an element size in the order of the vertical dislocation spacing h the density profile converges. The dislocation-dislocation interactions ensure a natural stabilisation, especially in the range of domain sizes considered here.

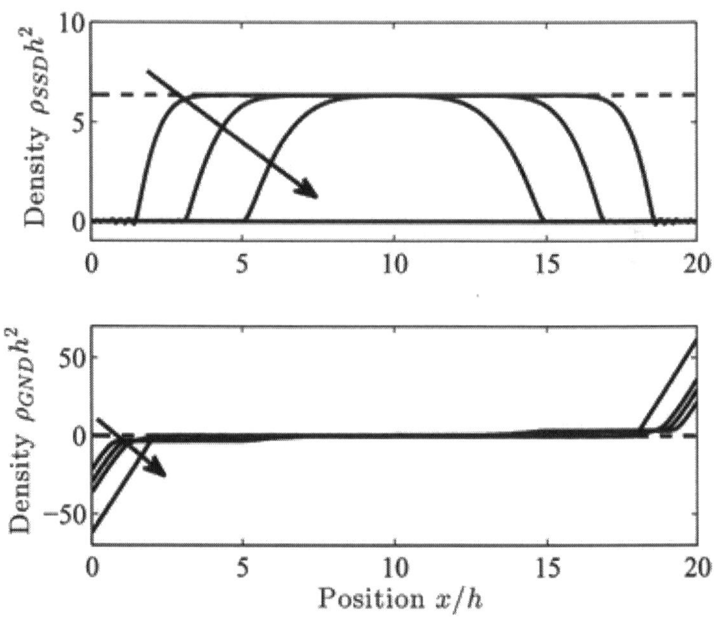

Figure 3 The evolution of an initially uniform dislocation density profile (*dashed lines*) containing both positive and negative walls under the influence of an applied shear stress. A double pile-up results, with positive and negative walls at opposite sides. Both the SSD (*top*) and GND (*bottom*) densities are shown

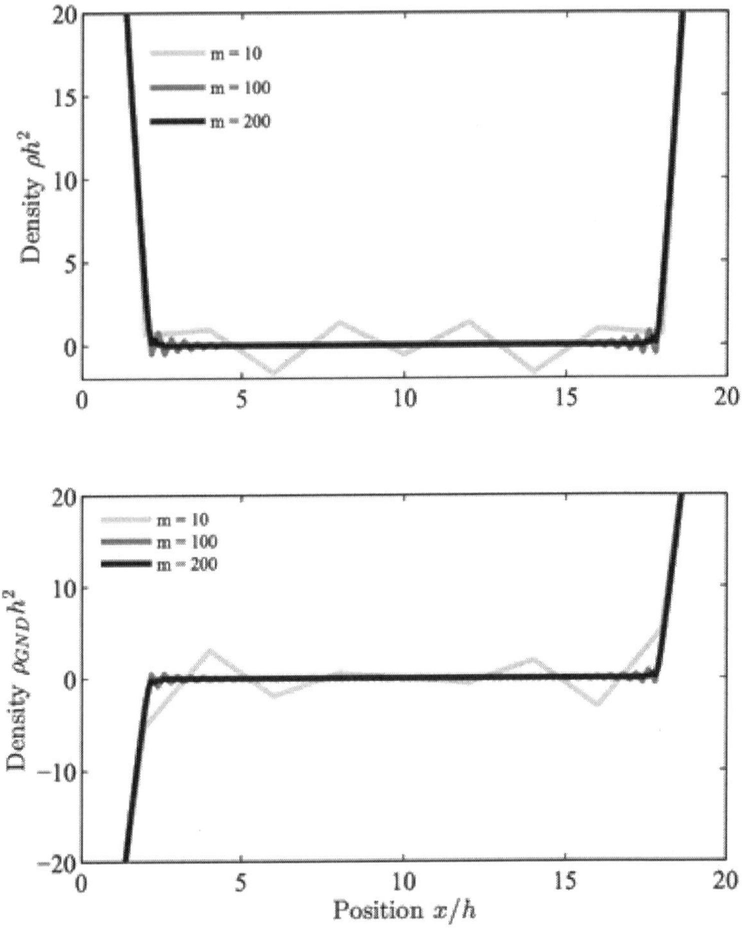

Figure 4 The influence of the number of elements m on the steady state dislocation density profiles of a double pile-up of positive and negative dislocations. Increasing in the number of elements results in convergence towards a (stable) solution

Next, the applied shear stress is removed, after which the accumulated interaction stresses are driving the dislocation walls back towards a uniform distribution. In Fig. 5, we can see the evolution of the dislocation density towards this equilibrium configuration.

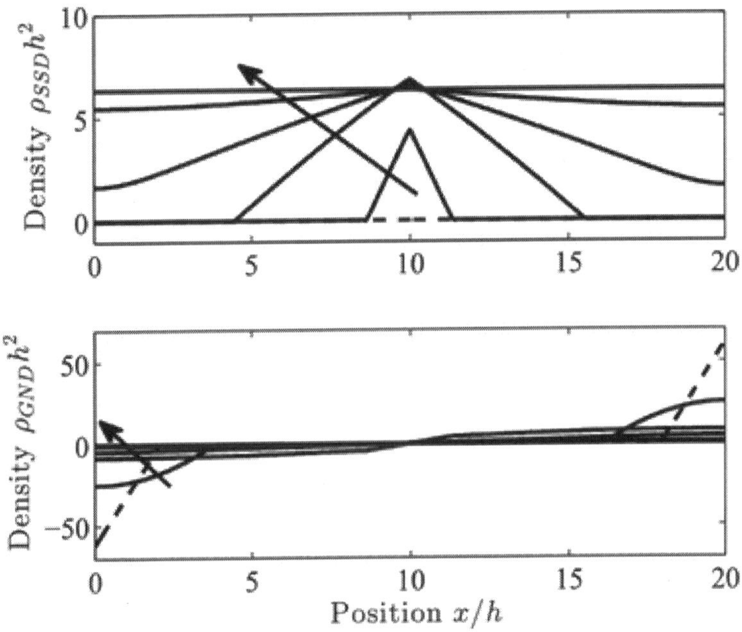

Figure 5 Evolution of the density after release of the stress. The *dashed line* indicates the initial density profile for this simulation, obtained after the pile-up of dislocations against the impenetrable boundaries. The dislocation walls recover their equilibrium positions

Figure 6 compares the final, equilibrium solutions of the pile-up and release cases with a discrete simulation, showing an adequate agreement. The continuum model captures the linear part of the double pile-up well and after release of the stress the equilibrium distribution is recovered quite accurately.

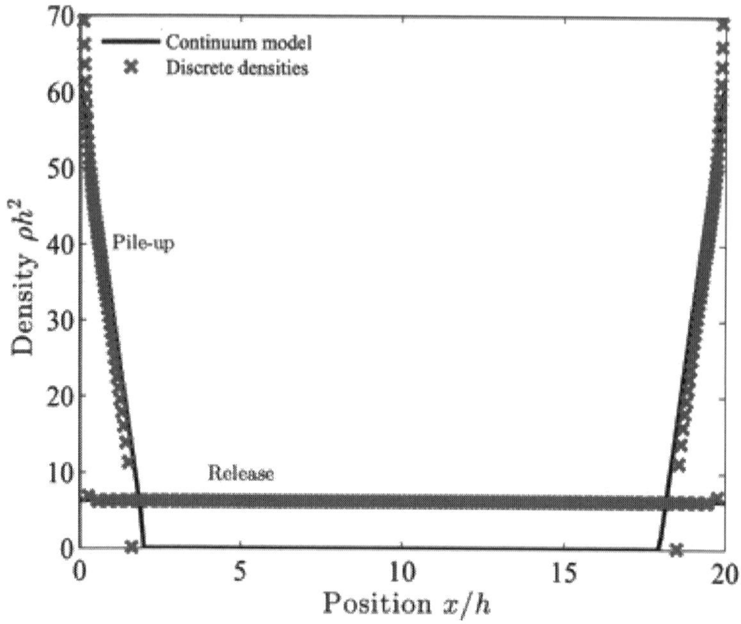

Figure 6 The steady state solutions of the total dislocation density for: (1) the double pile-up and (2) after the stress is released. The continuum solution is compared with a discrete simulation

The above simulations involved a large number of walls. With fewer dislocation walls, see in Fig. 7 for the case of 32 walls, the agreement between the discrete and continuum model is poorer when the stress is removed after the formation of a pile-up. For a smaller number of dislocations, the interaction stress is not high enough for the walls of opposing sign to pass each other and a different, non-uniform equilibrium solution results. Although the walls are redistributed, the positive and negative walls remain separated. This effect is not captured by the continuum model, where the densities return to their initial distribution. The use of first-order derivatives in the continuum expression for the interaction stress dictates that without an applied stress the dislocation velocity vanishes only when the gradients of the densities are zero. This triggers the uniformly distributed SSD density. A limitation of the continuum model therefore is that a sufficient number of dislocations must be present in order to accurately capture the behaviour of the dislocation walls. If not, the continuum assumption does not apply.

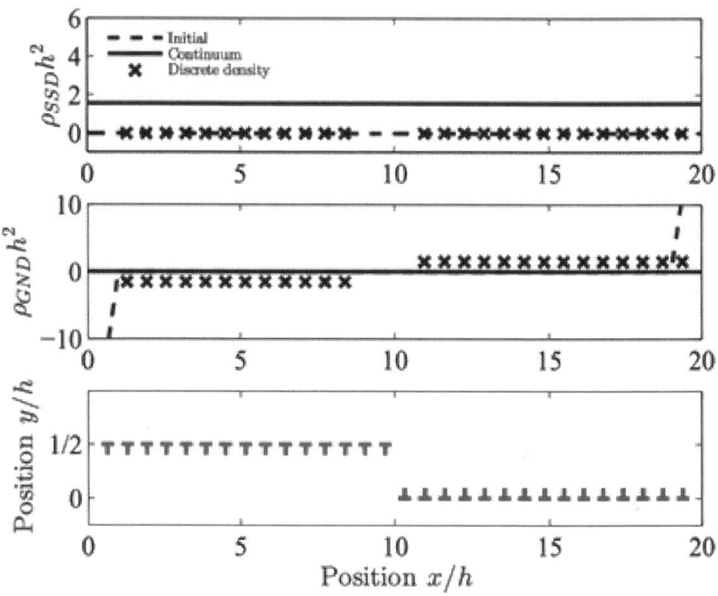

Figure 7 The steady state solutions of the total dislocation density (*top*) and the GND density (*middle*) when the stress is released after the double pile-up. In this case the number of dislocation walls is reduced to 32. The continuum solution is compared with a discrete simulation. The positions of the individual positive and negative walls are also shown (*bottom*)

4.2 Pile-up of positive walls

Now, we compare the new model with previous work and analyze its behaviour at the boundaries, in addition to the effect of the number of dislocation walls present. We study the case presented in [14] and [21] in which positive walls pile-up against a hard barrier, here located at x=0. This implies as boundary condition $\Phi\rho(x=0)=\Phi\kappa(x=0)=0$. Furthermore, we consider only positive dislocation walls in the domain. This corresponds with $\rho=\kappa,\rho SSD=0$in the continuum model.

The length of the domain is $L=10h$. First, a uniform distribution of positive walls is considered with an applied shear stress according to $\tau/\bar{G}=-1/20$. This results in a pile-up of dislocation walls for which the evolution is shown in Fig. 8. The applied stress pushes the dislocation walls towards the barrier, while at the same time the dislocation walls repel each other. This results in a pile-up against the barrier. The interaction stress emanates from the gradient of the density. In the final equilibrium state, this interaction stress must compensate the externally applied stress $-\tau$ which is constant. Therefore, a linear pile-up is predicted [21].

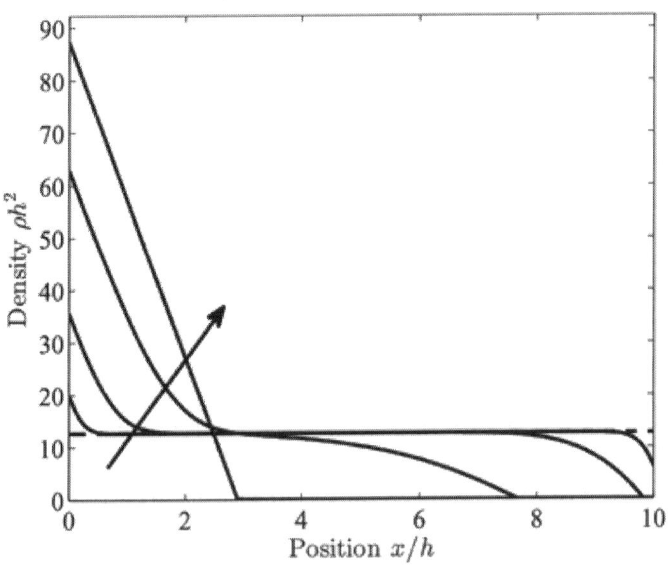

Figure 8 Pile-up of an initially uniform distribution of positive dislocations (*dashed line*) against an impenetrable barrier

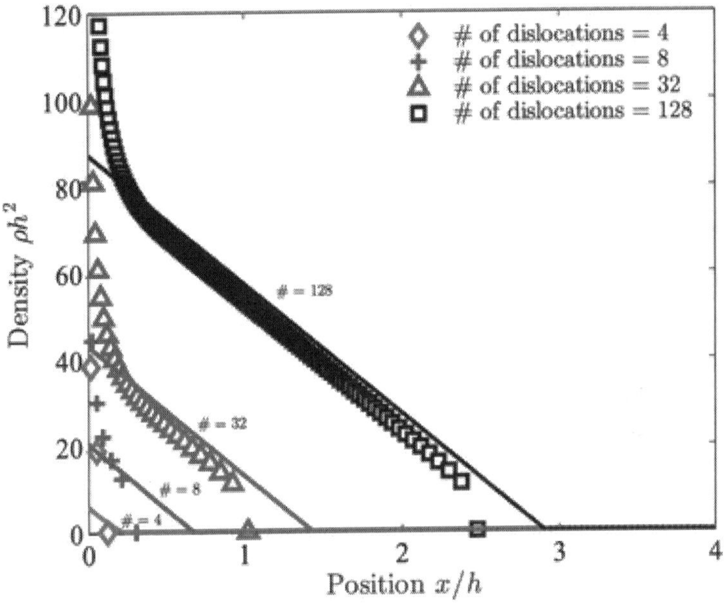

Figure 9 The pile-up of positive dislocation walls against an impenetrable barrier, predicted by the present continuum model and discrete dislocation simulations. The number of walls is varied

In Fig. 9, the equilibrium solution is shown for the present model and fully discrete simulations with different numbers of dislocations and thus initial densities. The results observed in Fig. 9 are consistent with the results found in [21]. The comparison between the continuum and discrete simulations is quite good for a large number of dislocations and at some distance from the barrier. However, if we decrease the number of walls below a certain limit, the continuum model no longer captures the discrete wall distribution, as the dislocations do not distribute linearly near the boundary. For all cases, the agreement in the region close to the boundary is poorer, because a boundary layer appears where the discrete nature of the dislocation walls is dominant. The assumptions made in deriving the continuum model do not hold in this region. In the expression for the velocity only the gradient of the density is related to the applied stress; therefore it is unable to capture the $\frac{1}{\sqrt{x}}$ behaviour near the obstacle. However, the linear part of the pile-up, as explained in [21], is captured accurately, i.e. the gradient of the density corresponds well between the discrete and continuum case. The absolute difference in density is a result of dislocation conservation in the domain. The area below the continuous and discrete data must be constant and equal. Because the continuum model is unable to capture the sharp peak near the boundary, the density in the region away from the boundary is higher.

4.3 Non-uniform distribution of positive and negative walls

To illustrate the effect of taking into account all short-range interaction stresses, and not only those between dislocations of the same sign, we consider equal numbers of positive and negative walls, corresponding in the continuum case with a distribution of SSDs, i.e. $\rho=\rho\mathrm{SSD},\kappa=0$. Initially, these walls are not horizontally equispaced, but their mutual distances are such that they result in a non-uniform density distribution as illustrated by the dashed curve in Fig. 10. In the center of our domain more walls are present, their number decreasing closer to the boundaries. The integral of the initial density profile corresponds with 89 dislocation walls. The boundaries are impenetrable barriers. No external stress is applied, such that the interactions between walls, due to the gradients in densities, constitute the only driving forces for dislocation motion. The length of the domain is $L=60h$.

In Fig. 10, the evolution of the SSD density is shown. The driving force of the dislocation density, i.e. the gradient in the density profile, results in mutually repelling walls and the spreading of the density distribution. The density profile is evolving to its equilibrium configuration, i.e. a uniformly distributed dislocation density.

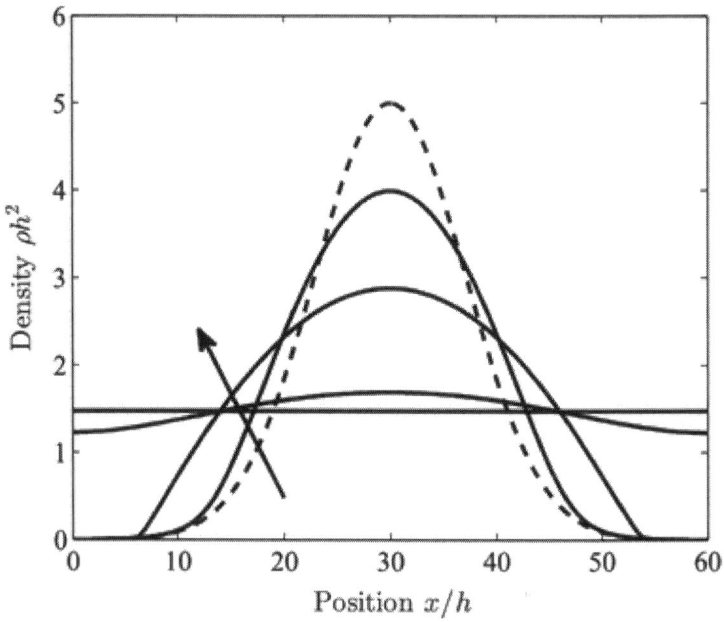

Figure 10 The evolution of an initially non-uniformly distributed SSD density (*dashed line*). Due to the repulsive interaction stresses, the density profile evolves to a uniformly distributed density

We next compare the present model to a discrete dislocation simulation as well as a continuum model including only GND interactions using the same number of walls, see Fig. 11.

In the present model (marked "all interactions") the dislocation walls tend to repel each other, thereby rearranging into a uniform distribution as is also observed in the discrete simulation. The offset between the discrete and the present continuum simulations is caused by the boundary condition in the discrete simulations. In the discrete simulation the impenetrable barrier is modelled by the addition of immobile dislocations. Two additional dislocation walls, one of each sign, are pinned at each boundary of the domain to ensure that no dislocations leave the domain. This results in an additional contribution to the interaction stress at the boundary and a slightly higher discrete density in the bulk.

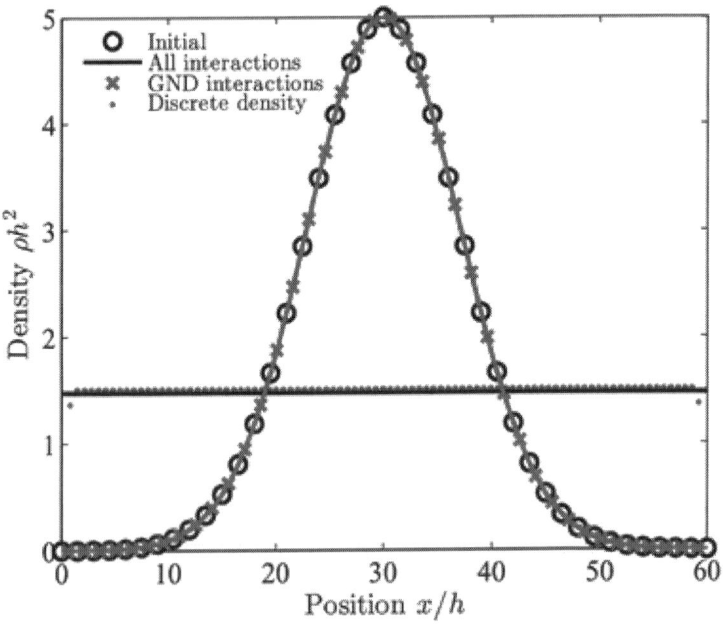

Figure 11 The steady state result of an initially non-uniformly distributed SSD density. The present model is here compared with discrete dislocation simulations and with a continuum model containing GND interactions only

The initial wall distribution (in which no GNDs are present) does not evolve for the model with only GND interactions. This model only includes the gradient of the GND density as a driving force for dislocation motion, which is obviously absent in the initial state. In the new, "all interactions" model the interactions between positive and negative walls is properly included via the gradient of the total dislocation density.

4.4 Outflow of dislocations

In the final example we replace the barrier at $x=0$ by a free outflow condition, regardless of the sign of the dislocations. A free boundary in the continuum framework is modelled by a Dirichlet condition on the densities, i.e. $\rho(0)=\kappa(0)=0$. Furthermore, no stress is applied and the length of the domain is set to $L=10h$. Our purpose is merely to illustrate the effect of accounting for all short-range dislocation interactions (as opposed to only GNDs) on the ability to model outflow. No comparison is therefore made with discrete simulations and only a qualitative analysis of the continuum models is performed.

Figure 12 shows a decrease in dislocation density near the boundary for the newly proposed model. The model with GND interactions only is unable to capture the outflow of dislocations, because there are no (gradients of) GNDs to generate the driving force in that formulation. In a qualitative sense the model presented here is thus capable of capturing dislocation outflow.

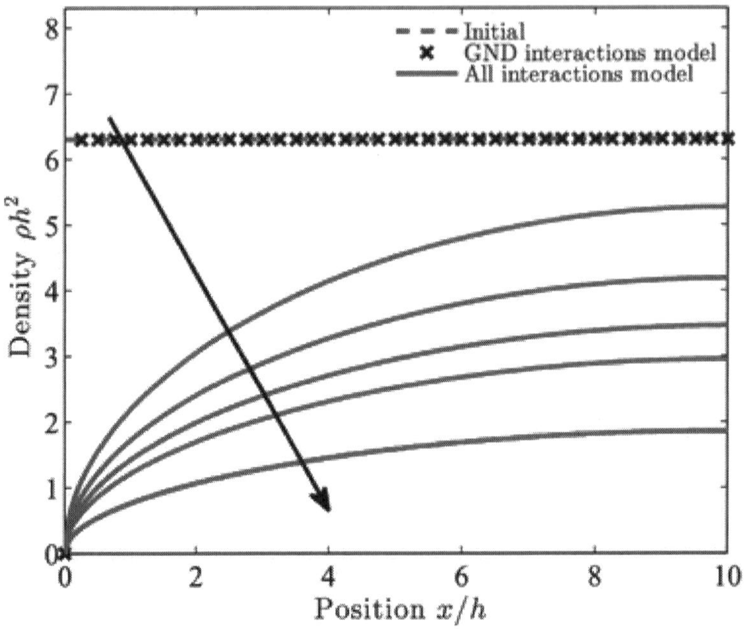

Figure 12 The evolution of the outflow of dislocations, initially uniformly distributed (*dashed line*), when a free boundary (at x=0) is present

5. CONCLUSION AND DISCUSSION

Starting from an idealised dislocation configuration, a continuum model for dislocation transport in single slip has been derived, including (short-range) dislocation-dislocation interactions. The driving force in the transport equations emanates from the behaviour of discrete dislocation walls. Whereas the long-range stresses in the driving force can be captured naturally in a continuum model, the short-range stresses are to be determined from the individual interactions between dislocations. These short-range stresses in the presented continuum model depend on the gradient of GND and total dislocation density, as opposed to other models found in literature in which they depend only on the gradient of GND density.

The resulting equations are non-linear transient PDEs. They are numerically solved by temporal and spatial discretisation and by adopting an incremental-iterative solution procedure. We develop a fully implicit scheme here, in contrast with the explicit time integration generally used in the literature. This framework was shown to convergence to a solution with a relatively low number of elements. In this study, no stabilisation is used for the transport equations, because this could interfere with the presence of the short-range dislocation interactions. When more realistic cases are considered, with larger problem length-scales, numerical stabilization may be necessary to avoid nonphysical solutions for tractable mesh sizes. Different approaches for stabilization of the dislocation transport equations as presented here (or similar) may be found in [22–24].

The behaviour of individual dislocation walls in a single slip system is captured adequately by our continuum model. However, a limitation of our model is its continuum nature, requiring a sufficient number of dislocation walls in the domain. A continuum description is unable to capture the distribution of dislocation walls when there is only a limited number of dislocations. The derivation of the interaction stress assumes that the walls are continuously distributed in the direction of the slip plane. In the case of a pile-up of walls against an impenetrable barrier, the linear part of the pile-up is captured accurately if the number of walls is sufficiently high. When the stress is removed in the case of a double pile-up, a continuum model cannot capture the discrete distribution when the number of dislocation walls is small.

It is shown that the short-range dislocation interactions can be easily implemented in an implicit finite element model for continuum dislocation transport. Although the current governing equations are formulated in terms of an idealised single-slip case, the ideas developed here may be extendable to more general cases with multiple active, and interacting, slip systems with mixed dislocations. As a first step, the extension to multiple slip systems with straight edge dislocations is considered in a forthcoming paper, assuming the Taylor-like organisation of the dislocations at the level of each individual slip system. This leads to balance laws for dislocation transport on each slip system which are equivalent with those for the single system considered here. The further extension to dislocations of mixed edge-screw type, cross-slip, etc. is a subject of further research.

NOTES

The authors declare that they have no competing interests.

The theory regarding continuum dislocation transport was derived by MD, with input and suggestions from RP and MG to define the scope of the study. The implementation of dislocation transport in the continuum model and obtaining the simulation results were carried out by MD. The manuscript was drafted by MD. RP helped to draft the manuscript by providing input during the various stages of the drafting proces. Both RP and MG also contributed by critically revising the manuscript regarding both text and intellectual content. All authors read and approved the final manuscript.

ACKNOWLEDGMENTS

This research was carried out under the project number M22.2.09342 in the framework of the Research Program of the Materials innovation institute M2i (http://www.m2i.nl).

REFERENCES

1. Roters F, Eisenlohr P, Hantcherli L, Tjahjanto DD, Bieler TR, Raabe D. Overview of constitutive laws, kinematics, homogenization and multiscale methods in crystal plasticity finite-element modeling: Theory, experiments, applications. Acta Mater. 2010;58(4):1152–211.

2. Lee EH. Elastic-plastic deformation at finite strains. J Appl Mech. 1969;36:1–6.MATH

3. Rice JR. Inelastic constitutive relations for solids: an internal variable theory and its application to metal plasticity. J Mech Phys Solids. 1971;19:433–55.MATH

4. Orowan E. Zur kristallplastizität. III: Über den mechanismus des gleitvorganges. Z Phys A-Hadron Nucl 1934; 89(9):634–659.

5. Nye JF. Some geometrical relations in dislocated crystals. Acta Metall Mater. 1953;1(2):153–62.

6. Ashby MF. The deformation of plastically non-homogeneous materials. Philos Mag. 1970;21:399–424.

7. Shu JY, Fleck NA. Strain gradient crystal plasticity: size-dependent deformation of bicrystals. J Mech Phys Solids. 1999;47(2):297–374.MATH

8. Gurtin ME. A gradient theory of single-crystal viscoplasticity that accounts for geometrically necessary dislocations. J Mech Phys Solids. 2002;50(1):5–32.MATHMathSciNet

9. Menzel A, Steinmann P. On the continuum formulation of higher gradient plasticity for single and polycrystals. J Mech Phys Solids. 2000;48(8):1777–96.MATHMathSciNet

10. Bargmann S, Ekh M, Runesson K, Svendsen B. Modeling of polycrystals with gradient crystal plasticity: a comparison of strategies. Philos Mag. 2010;90(10):1263–88.

11. Anand L, Gurtin ME, Lele SP, Gething C. A one-dimensional theory of strain-gradient plasticity: formulation, analysis, numerical results. J Mech Phys Solids. 2005;53(8):1789–826.MATHMathSciNet

12. Ekh M, Grymer M, Runesson K, Svedberg T. Gradient crystal plasticity as part of the computational modelling of polycrystals. Int J Numer Meth Eng. 2007;72(2):197–220.MATHMathSciNet

13. Evers LP, Brekelmans WAM, Geers MGD. Non-local crystal plasticity model with intrinsic SSD and GND Effects. J Mech Phys Solids. 2004;52(10):2379–401.MATH

14. Roy A, Peerlings RHJ, Geers MGD, Kasyanyuk Y. Continuum modeling of dislocation interactions: why discreteness matters? Mater Sci Eng. A. 2008;486(1–2):653–61.

15. Groma I, Csikor FF, Zaiser M. Spatial correlations and higher-order gradient terms in a continuum description of dislocation dynamics. Acta Mater. 2003;51(5):1271–81.

16. Yefimov S, Groma I, van der Giessen E. A comparison of a statistical-mechanics based plasticity model with discrete dislocation plasticity calculations. J Mech Phys Solids. 2004;52(2):279–300.MATHMathSciNet

17. Geers MGD, Peerlings RHJ, Hoefnagels JPM, Kasyanyuk Y. On a proper account of first- and second-order size effects in crystal plasticity. Adv Eng Mater. 2009;11(3):143–7.

18. Cleveringa HHM, van der Giessen E, Needleman A. Comparison of discrete dislocation and continuum plasticity predictions for a composite material. Acta Mater. 1997;45(8):3163–79.

19. Arsenlis A, Parks DM. Modeling the evolution of crystallographic dislocation density in crystal plasticity. J Mech Phys Solids. 2002;50(9):1979–2009.MATH

20. Hirth JO, Lothe J. Theory of Dislocations. Florida: Krieger Publishing Company; 1992.

21. de Geus TWJ, Peerlings RHJ, Hirschberger CB. An analysis of the pile-up of infinite periodic walls of edge dislocations. Mech Res Commun. 2013;54:7–13.

22. Varadhan SJ, Beaudoin AJ, Acharya A, Fressengeas C. An analysis of the pile-up of infinite periodic walls of edge dislocations. Modelling Simul Mater Sci Eng. 2006;14:1245–70.

23. Kords C. On the role of dislocation transport in the constitutive description of crystal plasticity. PhD thesis, Rheinish-Westfälischen Technischen Hochshule Aachen, 2013.

24. Hernández H, Massart TJ, Peerlings RHJ, Geers MGD. An analysis of the pile-up of infinite periodic walls of edge dislocations. Modelling Simul Mater Sci Eng. 2015;23:085013.

Index